A Guide to Infinity

Edward R. Scheinerman

A GUIDE TO INFINITY
Ten Mathematical Journeys

Yale University Press
New Haven and London

Copyright © 2026 by Edward R. Scheinerman.
All rights reserved.

This book may not be reproduced, in whole or in part, including illustrations, in any form (beyond that copying permitted by Sections 107 and 108 of the U.S. Copyright Law and except by reviewers for the public press), without written permission from the publishers.

Yale University Press books may be purchased in quantity for educational, business, or promotional use. For information, please e-mail sales.press@yale.edu (U.S. office) or sales@yaleup.co.uk (U.K. office).

Printed in the United States of America.

Library of Congress Control Number: 2025937438
ISBN 978-0-300-28479-9 (hardcover)

A catalogue record for this book is available from the British Library.

Authorized Representative in the EU: Easy Access System Europe, Mustamäe tee 50, 10621 Tallinn, Estonia, gpsr.requests@easproject.com

10   9   8   7   6   5   4   3   2   1

To Carmella, Shlomit, Gershom, Raya, Zev, and Isaac

Contents

Preface   xi

Part I: Real Infinity   1

1 Real Numbers, Extended   3
    1.1 Real numbers: Operations and properties   4
    1.2 Two new numbers: ∞ and $-\infty$ 7
    1.3 Computing with ∞ and $-\infty$ 11
    For your consideration 13

2 Infinite Decimals   15
    2.1 Place value notation 15
    2.2 What is $0.3333\ldots$? 16
    2.3 Infinite sums as sequences 17
    2.4 Smaller terms but infinite sum 19
    For your consideration 21

3 Tropical Arithmetic   23
    3.1 New operations: $\oplus$ and $\odot$ 23
    3.2 Polynomials 26
    3.3 Tropical algebraic curves 28
    For your consideration 33

4 Hyperreals   35
    4.1 Starting small 35
    4.2 The transfer principle 37
    4.3 Derivatives made easy 41
    For your consideration 43

Part II: Plane Infinity   45

5   Complex Numbers, Extended   47
- 5.1 Complex numbers 47
- 5.2 One infinity for all 49
- 5.3 Stereographic projection and the Riemann sphere 50
- 5.4 Linear fractional transformations 52
- For your consideration 56

6   Line at Infinity I: Projective Plane   59
- 6.1 Parallel lines meet at infinity 59
- 6.2 The projective plane 60
- 6.3 Points and lines, algebraically 63
- 6.4 Desargues's theorem 67
- 6.5 Topology of the projective plane 70
- 6.6 Bézout's theorem 74
- For your consideration 80

7   Line at Infinity II: Hyperbolic Plane   83
- 7.1 The parallel postulate 83
- 7.2 The hyperbolic plane 85
- 7.3 Tessellations 92
- 7.4 Distance and area in the hyperbolic plane 97
- For your consideration 100

Part III: Infinite Things   101

8   Counting Infinity   103
- 8.1 Counting things 103
- 8.2 Sets, functions, bijections 104
- 8.3 Finite and infinite sets 110
- 8.4 Comparing infinite sets 113
- 8.5 Transfinite cardinal numbers 118
- 8.6 Transcendental numbers 123
- For your consideration 125

9   Order Infinity   127
- 9.1 Ordered sets 127
- 9.2 Well-ordered sets 129
- 9.3 Ordinal numbers 133

For your consideration 139

10 Infinite Shapes **141**
　　　10.1 Circles 141
　　　10.2 How long is the diagonal of a square? 143
　　　10.3 Sierpiński's carpet 145
　　　10.4 Hilbert's space-filling curve 151
　　　10.5 Mandelbrot set 156
　　　For your consideration 162

Bibliography **167**

Index **171**

Preface

> Nevertheless, as an individual, I may be permitted to say that I cannot conceive Infinity, and am convinced that no human being can.
> —Edgar Allan Poe, *Eureka, A Prose Poem*

Infinity is a mind-boggling concept. It has been the subject of philosophers, physicists, theologians, and poets. One might think that trying to grasp the notion of infinity is, as Edgar Allan Poe asserts, inconceivable. It may indeed be beyond human thought to fully comprehend immortality or our infinitesimal significance in the universe.

However, for mathematicians, working with infinity is just another day at the office.

The goal of this book is to introduce readers to a variety of ways in which mathematicians think about and use infinity.

Infinity is not one thing. Infinity can simply be a number adjoined to the real or complex numbers. It can also be the answer to a counting problem such as "How many even integers are there?" In this latter case, two different counting problems—both with infinite results—can have different infinite answers. Infinity is also a faraway place where parallel lines meet.

Infinity is not inscrutable: What you need to know

The infinite can be terrifying; see the passage by James Joyce, "A description of infinite time," on page xiii.

Mathematical infinity, however, is no more ferocious than -1, or $\sqrt{2}$, or π. Our presentation of the infinite, in its many guises, only requires mathematical fluency at the pre-calculus level. The tools we employ are algebra, complex numbers, geometry, and basic logic.

The twentieth-century Hungarian-American mathematician John von Neumann quipped that "in mathematics you don't understand things; you just get used to them." This applies to mathematical infinity. We can work comfortably with the infinite without ever having to grasp its enormity.

Overview

This book is divided into three parts.

Part I is about *Real Infinity*. These chapters deal with infinity in the context of the real number system. We begin (Chapter 1) with simply adjoining two new numbers, ∞ and −∞, to the reals and see where that leads us. We also take a careful look (Chapter 2) at infinite decimal numbers. Next we introduce tropical arithmetic (Chapter 3), in which the usual operations of addition and multiplication are replaced in a way that naturally incorporates infinity. Finally, the "crude" addition of ±∞ to the reals gives them a massive "sophisticated upgrade" to yield the hyperreals (Chapter 4), which incorporate infinitesimally small numbers and an infinity of infinite values.

In Part II, *Plane Infinity*, we approach infinity geometrically. Whereas in Chapter 1 we extended real numbers with the addition of ∞ and −∞, in Chapter 5 we extend the complex numbers with a single infinity. In so doing, we visualize the extended complex numbers as a sphere. Next we think about parallel lines. In Chapter 6 we augment the usual Euclidean plane by imagining that parallel lines actually meet—at infinity. Incorporating a line at infinity transforms the standard Euclidean plane into the projective plane. We learn that the line at infinity isn't any different from any other line. Parallel lines are important again in Chapter 7 where we learn about the hyperbolic plane. Although infinite in size, the entire hyperbolic plane can be visualized inside a circle. There is a beautiful consequence of this alternative geometry: The hyperbolic plane can be tiled with regular polygons in infinitely many beautiful ways.

Part III, *Infinite Things*, is concerned with infinite objects. Natural numbers are descriptors of *finite* sets; they specify the number of members held in the set. In Chapter 8 we extend this idea to *infinite* sets where we encounter the transfinite cardinal numbers. This chapter includes the celebrated work of Georg Cantor, who showed that infinite sets are not all the same size. Sets are unordered collections, but in Chapter 9 we explore transfinite ordinal numbers that are descriptors of (a special type of) ordered sets. Arithmetic with transfinite ordinals is surprising because

A description of infinite time

You have often seen the sand on the seashore. How fine are its tiny grains! And how many of those tiny little grains go to make up the small handful which a child grasps in its play. Now imagine a mountain of that sand, a million miles high, reaching from the earth to the farthest heavens, and a million miles broad, extending to remotest space, and a million miles in thickness; and imagine such an enormous mass of countless particles of sand multiplied as often as there are leaves in the forest, drops of water in the mighty ocean, feathers on birds, scales on fish, hairs on animals, atoms in the vast expanse of the air: and imagine that at the end of every million years a little bird came to that mountain and carried away in its beak a tiny grain of that sand. How many millions upon millions of centuries would pass before that bird had carried away even a square foot of that mountain, how many eons upon eons of ages before it had carried away all? Yet at the end of that immense stretch of time not even one instant of eternity could be said to have ended. At the end of all those billions and trillions of years eternity would have scarcely begun. And if that mountain rose again after it had been all carried away, and if the bird came again and carried it all away again grain by grain, and if it so rose and sank as many times as there are stars in the sky, atoms in the air, drops of water in the sea, leaves on the trees, feathers upon birds, scales upon fish, hairs upon animals, at the end of all those innumerable risings and sinkings of that immeasurably vast mountain not one single instant of eternity could be said to have ended; even then, at the end of such a period, after that eon of time the mere thought of which makes our very brain reel dizzily, eternity would scarcely have begun.
—James Joyce, *A Portrait of the Artist as a Young Man*

addition is not commutative. Finally, in Chapter 10 we explore some examples of infinite shapes. We begin by imagining circles as regular polygons with infinitely many sides. We also meet fractals and space-filling curves.

Every chapter concludes with some thought-provoking questions for the reader's consideration.

The sequential dependence of the chapters is modest. While we recommend that the chapters be read in the order presented, it is safe to follow a path that piques your interest.

Acknowledgments

The motivation for writing about infinity came from my granddaughter Shlomit, who asked me: *What is the number just before infinity?* Although the answer is disappointing (in the context of the extended real numbers, there is no such number), the question is marvelous. Thank you, Shlomit!

I want to thank friends and colleagues who gave me feedback on early versions of this book. Thank you to: Pamela Franks, Isaac Kamionkowski, Evan MacMillan, John MacMillan, Joseph McCloskey, Evelyne Rottiers, James Schatz, Rebecca Schulman, and Daniel Velleman.

Thanks also to anonymous reviewers for their thoughtful suggestions that certainly have made this book better.

And many, many thanks to Yale University Press, especially my editor Jean Black for her enthusiasm and wonderful discussions that improved the book. Thank you to her editorial assistant Elizabeth Sylvia, not only for her technical help on a variety of logistical matters, but also for her suggestion for this book's title. Thank you to Phillip King and Eva Skewes for all their work to bring this project to fruition. Finally, I am deeply grateful to MaryEllen Oliver for her meticulous and thoughtful copyediting.

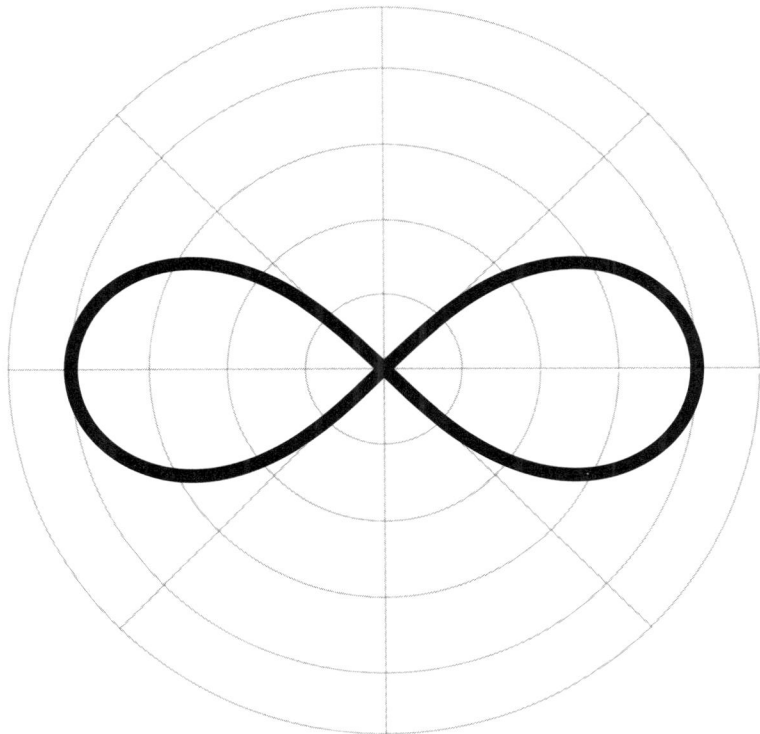

This curve, called a lemniscate, is the polar plot of the equation $r^2 = \cos(2\theta)$. We illustrate polar coordinates to represent complex numbers as a point in the plane in Figure 5.4 (on page 54).

Part I: Real Infinity

Chapter 1

Real Numbers, Extended

> You can't have everything. Where would you put it?
> —Steven Wright

We begin by introducing ∞ as a number. Numbers are often visualized as points on a line—a number line. As the line goes forever to the left and to the right, where would we put it? Further, how does the new number ∞ interact with the ordinary numbers (using the usual operations such as addition and multiplication)? Ideally, all the usual rules of algebra (such as the associative property) should remain intact once we extend operations to include ∞. Alas, as we shall see, we can't have everything.

The incorporation of new numbers into mathematics has a long history. Millennia ago, humans conceived of numbers for counting: the familiar 1, 2, 3, and so on. Later 0 was added to the family of numbers, with a bit of controversy.[1] Fractions are comfortable extensions; negative numbers less so.

Why, we ask, is the product of negative numbers positive? Wouldn't multiplying two negative numbers be "even more negative" than either factor?

It was a *decision* to extend the definition of multiplication to negative numbers so that $-2 \times -3 = 6$ and not -6 (or anything else). Why that decision? The answer is that we want the basic properties of addition and multiplication (such as the commutative property and the distributive

---

[1] See Charles Seife's *Zero: The Biography of a Dangerous Idea*.

property) to continue to hold for all numbers, not just the positive numbers. The decision to ensure the validity of the basic properties of + and × forces our hand.[2]

We are now poised to welcome ∞ as a number that is greater than all (ordinary) real numbers. To preserve algebraic properties, what decisions do we need to make?

1.1 Real numbers: Operations and properties

Mathematics is teeming with all kinds of numbers of which *real numbers* are the most familiar.

Real numbers are typically written in decimal (base ten) notation using the usual digits (0 through 9), often a decimal point, sometimes starting with a minus sign, and occasionally with commas to improve readability. Here are some examples:

$$1 \quad 2 \quad -3 \quad 1.25 \quad 12{,}345 \quad 0 \quad 3.141592653\ldots$$

The set of all real numbers is denoted by $\mathbb{R}$.

Real numbers represent finite quantities. In this chapter, we extend the real numbers to include infinite quantities.

Despite their name, real numbers aren't *real*. (See "Can you just make up numbers?" on page 8.) They are simply mental constructs. Real numbers are *useful* because they model real-world phenomena such as distances, densities, debts, durations, and so forth. They become even more useful and interesting when we use them in calculations and comparisons.

The most basic calculations for real numbers are performed with the operations of addition (+) and multiplication (·), as well as their cousins, subtraction and division. Besides equality, the usual way to compare real numbers is with the less-than relation (<) and its cousin, greater-than.

Before we modify the real number system to include infinity, we recall the basic properties of operations and comparisons.

---

[2]The details of this are not difficult, but would be a distraction from our journey. Try to work this out for yourself starting with the facts that $-x + x = 0$ and that anything multiplied by 0 is 0.

Arithmetic properties of the real numbers

Addition and multiplication of real numbers obey fundamental properties such as the commutative, associative, and distributive properties. Let's review what those are.

Given any two real numbers x and y, the result of adding or multiplying these numbers does not depend on the order in which they appear in the calculation. That is, $x + y$ and $y + x$ are always equal, as are $x \cdot y$ and $y \cdot x$. When order doesn't matter, we say that the operation is *commutative*.

Commutative Property for Addition and Multiplication
For any two real numbers, x and y, we have $x + y = y + x$ and $x \cdot y = y \cdot x$.

Another order-doesn't-matter property for addition and multiplication involves triples of numbers. When we add three numbers, x, y, and z, it doesn't matter if we first add x to y, and then add z, or if we first add y and z, and then add x. The same is true for multiplication. This is known as the *associative* property.

Associative Property for Addition and Multiplication
For any three real numbers, x, y, and z, we have $(x + y) + z = x + (y + z)$ and $(x \cdot y) \cdot z = x \cdot (y \cdot z)$.

The commutative and associative properties pertain to single operations. The *distributive* property involves both addition and multiplication.

Distributive Property
For any three real numbers, x, y, and z, we have $x \cdot (y + z) = x \cdot y + x \cdot z$.

Both addition and multiplication have *identity elements*, 0 and 1, respectively.

Identity Elements for Addition and Multiplication
Let x be any real number. Then $0 + x = x$ and $1 \cdot x = x$.

The operations of subtraction and division are built on addition and multiplication, respectively, by way of *inverses*. For a real number x, its *additive inverse* is a number y so that $x + y = 0$. In other words, y is simply the *negative* of x: $y = -x$. Similarly, for a nonzero real number x, its *multiplicative inverse* is a number y such that $x \cdot y = 1$. The multiplicative inverse of x is also known as the *reciprocal* of x, and the notation is x^{-1}.

> **Operation Inverses**
> Let x be any real number. Then there is a real number $-x$ such that $x + (-x) = 0$.
> If x is nonzero, then there is a real number x^{-1} such that $x \cdot x^{-1} = 1$.

Note that there is no reciprocal of zero. This is because $0 \cdot x = 0$ for any real number x; there is no way to multiply 0 times something and get 1 as a result.

The operations of subtraction and division are defined by way of addition, multiplication, and inverses. Specifically:

$$x - y = x + (-y) \quad \text{and} \quad x \div y = x \cdot y^{-1}.$$

Division by zero is not allowed because there is no reciprocal of 0. Stated differently, division by zero is *undefined*, which simply means that we do not ascribe any value to $x \div 0$.

Order properties of real numbers

Operations, such as addition and multiplication, *combine* real numbers to produce a real number result. Real numbers can be *compared* using the familiar less-than relation, <.

The less-than relation has two important properties. First, if $x < y$ and $y < z$, then we may conclude that $x < z$. This is known as the *transitive* property.

> **Transitive Property**
> Let x, y, and z be real numbers. If $x < y$ and $y < z$, then it must be the case that $x < z$.

Any two real numbers are either equal to each other, or else one is less than the other. This is formalized as the *trichotomy* property.

Trichotomy Property
  Let x and y be real numbers. Then exactly one of the following is true: $x < y$, or $x = y$, or $y < x$.

Notice that the relations $\leq$, $>$, and $\geq$ may all be defined from the basic $<$ relations:

- $x \leq y$ means either $x = y$ or $x < y$.

- $x > y$ means $y < x$.

- $x \geq y$ means $x = y$ or $y < x$.

Arithmetic and order properties combined

We now consider how the arithmetic properties of addition and multiplication relate to the ordering of real numbers by the less-than relation.

Recall that a real number x is called *positive* if $x > 0$ and *negative* if $x < 0$.

Operations and Order
  Let x and y be real numbers.

- If x and y are both positive, so are $x + y$ and $x \cdot y$.

- If x and y are both negative, then $x + y$ is negative and $x \cdot y$ is positive.

- If one of x and y is positive, and the other is negative, then $x \cdot y$ is negative.

1.2   Two new numbers: ∞ and $-\infty$

With this review in hand, we are ready to extend the real number system by introducing two new numbers: ∞ and $-\infty$. Simply naming two new numbers is insufficient; we need to see how they relate and interact with (ordinary) real numbers.

> **Can you just make up numbers?**
>
> All numbers are made up; they are purely mental creations. There is no such "thing" as the number 5 or the number −2 or any other number. To be sure, some numbers are so natural that they seem inevitable, especially the *natural numbers*, 0, 1, 2, 3, and so forth.
>
> Just because we call numbers such as $\frac{2}{3}$ or $\sqrt{2}$ *real* does not make them real! In Chapter 5, we consider the complex numbers; the fact that we call numbers such as $i = \sqrt{-1}$ *imaginary* does not mean that they are any more abstract or any less substantial than the real numbers.
>
> May you, my reader, make up numbers of your own? Certainly. You may name your new numbers Fred or Ginger or whatever you like. The question becomes: Are your new numbers *interesting* or *useful* (or both)?
>
> Just saying that you have a new number named Hermione is neither interesting nor useful. Can you describe how Hermione works with other numbers (say, with the real numbers or with other new numbers you've created named Ron and Harry)? What interesting properties do your numbers have? Do they model any real-world phenomena?

When ∞ and $-\infty$ are incorporated with the real numbers, we call the resulting system the *extended real numbers*, and we use the symbol $\overline{\mathbb{R}}$ to stand for the set that includes all the real numbers as well as ∞ and $-\infty$. In symbols, $\overline{\mathbb{R}} = \mathbb{R} \cup \{-\infty, \infty\}$.

Having named the two new numbers, we extend the operations (addition, subtraction, multiplication, division) and the less-than relation to include ∞ and $-\infty$. We set as a priority not to violate the various properties in Section 1.1.

Ordering

First of all, ∞ should be big! Indeed, it should be greater than any real number. Likewise, $-\infty$ should be infinitely negative; it should be less than any real number. In symbols, if x is any real number, we have

$$-\infty < x < \infty. \tag{1.1}$$

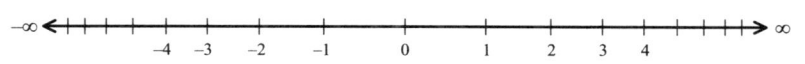

Figure 1.1: Visualizing the extended real number line, $\overline{\mathbb{R}}$, formed by appending $-\infty$ and ∞ to the real numbers.

It is helpful to have a way to visualize $\overline{\mathbb{R}}$. The real numbers $\mathbb{R}$ can be depicted as a number line. This representation sees the real numbers displayed horizontally, as a line in which each point on the line represents a real number. To visualize $\overline{\mathbb{R}}$, we simply place ∞ all the way to the right and $-\infty$ all the way to the left, as in Figure 1.1.

Equation (1.1) nicely captures how ∞ and $-\infty$ relate to each other and to the real numbers. It allows us to comfortably say that ∞ is positive and $-\infty$ is negative.

The next step is to determine, to the maximum extent possible, how arithmetic operations may be extended to $\overline{\mathbb{R}}$.

Addition and subtraction

For a real number x, how shall we define $x + \infty$? Quite reasonably, we choose the result to be ∞. Likewise, for a real number x, we have $x + -\infty = -\infty$.

We want our extended definition of addition to obey the commutative property, so we also decide that $\infty + x = \infty$ and $-\infty + x = -\infty$ for any real number x.

What value shall we give to $\infty + \infty$? The logical choice is $\infty + \infty = \infty$. Likewise, $-\infty + -\infty = -\infty$.

We have covered nearly all possibilities except this: What value should we assign to $-\infty + \infty$?

Let's rule out $-\infty + \infty = -\infty$ and $-\infty + \infty = \infty$. There's no sense in which either ∞ or $-\infty$ is "stronger" than the other. So that suggests we choose a finite value for $-\infty + \infty$. A compelling choice is $-\infty + \infty = 0$.

Alas, this causes problems. Consider this calculation: $-\infty + \infty + \infty$. What's the answer? We want to preserve the associative property, which leads to this conundrum:

$$(-\infty + \infty) + \infty = 0 + \infty = \infty$$
$$\text{but} \quad -\infty + (\infty + \infty) = -\infty + \infty = 0.$$

So either we abandon the associative property or we can't have $-\infty + \infty$ equal to 0. Unfortunately, any other choice for $-\infty + \infty$ leads to exactly the same problem.

We have a choice: Either we abandon the associative property or we decide that $-\infty + \infty$ is undefined. The convention chosen by mathematicians is to leave this as undefined.

Despite the fact that $-\infty + \infty$ is undefined, we do consider ∞ and $-\infty$ to be negatives of each other. That is,

$$-(\infty) = -\infty \quad \text{and} \quad -(-\infty) = \infty.$$

Alas, $-\infty$ and ∞ are not additive inverses of each other, but the notation is not unreasonable.

Subtraction can now follow the usual definition—namely, that $x - y$ is defined as $x + (-y)$ regardless of where x (or y) are real, $-\infty$, or ∞. This gives rise to results such as $x - \infty = -\infty$ and $x - (-\infty) = \infty$ (where x is a real number). However, $\infty - \infty$ is undefined.

Multiplication and division

Next we extend multiplication for $\overline{\mathbb{R}}$. Let's begin with multiplying $\infty \cdot \infty$. Quite reasonably, we set this to ∞. Likewise, $\infty \cdot (-\infty) = (-\infty) \cdot \infty = -\infty$ and $(-\infty) \cdot (-\infty) = \infty$.

Next we consider the result of multiplying ∞ by a real number.

If x is a positive real number, we have $x \cdot \infty = \infty$. This comports with our understanding of multiplication as repeated addition. Adding a positive number, x, to itself infinitely many times should yield ∞.

Likewise, if x is positive, we set $x \cdot -\infty = -\infty$.

Finally, if x is negative, we have $x \cdot \infty = -\infty$ and $x \cdot -\infty = \infty$.

We have a decision to make in defining $0 \cdot \infty$. Unfortunately, there are only bad choices.

Perhaps we can set $0 \cdot \infty = 1$? If that were the case, we would have $1 = 0 \cdot \infty = (0 + 0) \cdot \infty = 0 \cdot \infty + 0 \cdot \infty = 1 + 1 = 2$ (by the distributive property).

Might we define $0 \cdot \infty = 0$? Not a terrible choice, but why should zero overpower infinity in this multiplication? It is just as sensible to say that $0 \cdot \infty = \infty$.

The bottom line is that mathematicians prefer not to assign a value to $0 \cdot \infty$ and leave it undefined.

Next we consider division. When we divide real numbers x and y, we are imagining x chopped into y equal-sized pieces. If y is very large, the pieces are very small. For this reason, we define $x \div \infty = 0$ for real numbers x. (Likewise, $x \div -\infty = 0$.)

Placing ∞ in the numerator is not problematic, so long as the denominator is a nonzero real number. For example, if x is positive, then $\infty \div x = \infty$. This is tantamount to $\infty \cdot x^{-1}$, which we have already decided is ∞. Dividing ∞ by a negative real number gives $-\infty$.

Trouble brews when numerator and denominator are both infinite. The convention is that $\infty \div \infty$ is undefined.

Finally, we reach the pesky issue of division by 0. It is reasonable to define $x \div 0 = \infty$ whenever x is positive and $x \div 0 = -\infty$ whenever x is negative. However, there is no reasonable way to assign a value to $0 \div 0$; the convention is to leave that undefined.

1.3 Computing with ∞ and $-\infty$

The inclusion of ∞ and $-\infty$ with real numbers gives rise to the extended real number system $\overline{\mathbb{R}}$. The basic operations (addition, subtraction, multiplication, and division) for real numbers are partially, but not fully, interoperable with ∞ and $-\infty$. A summary of these operations for the extended real numbers is summarized in Figure 1.2.

Computing with infinity is more than a theoretical exercise. Many computer languages support calculations involving ordinary real numbers (called *floating-point* numbers in computer lingo) and infinity.

The Institute of Electrical and Electronics Engineers has published a set of guidelines for how computers should handle expressions that involve infinity; these guidelines are known as the *IEEE Standard for Floating-Point Arithmetic* (IEEE Standard 754).

In the Python language, one can set a variable to represent infinity using this statement: `inf = float('inf')`. The variable `inf` can then be used with ordinary numbers. Some operations result in the value `nan`.

$$-\infty < x < \infty$$

$$x + \infty = \infty$$

$$\infty + \infty = \infty$$

$$\infty - \infty \quad \text{undefined}$$

$$\infty \cdot \infty = \infty$$

$$x \cdot \infty = \begin{cases} \infty & x > 0 \\ -\infty & x < 0 \\ \text{undefined} & x = 0 \end{cases}$$

$$x \div \infty = 0$$

$$\infty \div \infty \quad \text{undefined}$$

$$\infty \div 0 = \infty$$

$$x \div 0 = \begin{cases} \infty & x > 0 \\ -\infty & x < 0 \\ \text{undefined} & x = 0 \end{cases}$$

Figure 1.2: Summary of relations and operations in $\overline{\mathbb{R}}$ involving ∞ and a real number x. Operations with $-\infty$ are analogous.

This symbol stands for *not a number* and corresponds to our declaration that certain operations are undefined.

Here is a sample Python session in which we illustrate some calculations involving ∞. Lines that begin with >>> contain user input; the other lines are the output from the computer.

```
>>> inf = float('inf')
>>> inf + 1
inf
>>> 2*inf
inf
>>> 5-inf
-inf
>>> 1/inf
0.0
>>> inf-inf
nan
>>> inf/inf
nan
```

Division by 0 results in an error in Python:

```
>>> 1.0 / 0.0
Traceback (most recent call last):
  File "<stdin>", line 1, in <module>
ZeroDivisionError: float division by zero
```

Some other computer languages allow division by zero and give a result standing for infinity. For example, the following program is written in C++:

```
#include <iostream>
using namespace std;

int main() {
    // compute 1÷0 and print the result
    cout << 1.0 / 0.0 << endl;
}
```

When compiled and run, the output is `inf`.

For your consideration

When we append ∞ and $-\infty$ to the reals, we need to decide how they work with the usual operations of addition, subtraction, multiplication, and division. In so doing, we give priority to core algebraic properties, such as the associative property. For example, we are comfortable defining $\infty + \infty = \infty$ both because it makes sense intuitively and also because this definition does not violate any algebraic properties. However, any value we might ascribe to $\infty - \infty$ leads to a violation of an algebraic property, and so we choose to leave that expression undefined.

We have not explored how ∞ and $-\infty$ work with exponentiation. How would you define expressions such as these?

$$2^{\infty}, \quad 2^{-\infty}, \quad \infty^{\infty}, \quad \infty^{-\infty}, \quad 0^{\infty}, \quad \text{and} \quad \infty^{0}$$

As you consider how to define each of these (or leave some undefined), an algebraic property to preserve is the following:

$$\left(a^{b}\right)^{c} = a^{bc}$$

where $a \geq 0$.

Chapter 2

Infinite Decimals

> ... it's very much like your trying to reach infinity. You know that it's there, you just don't know where—but just because you can never reach it doesn't mean that it's not worth looking for.
>
> —Norton Juster, *The Phantom Tollbooth*

Divide 1 by 3 on a calculator and the result is not exact; the display shows 0.3333333, which is nearly $\frac{1}{3}$, but we can never reach it. To get an exact result, we need to go to infinity; it's worth the trip.

2.1 Place value notation

Real numbers are typically written in decimal notation. This notation is also called *place value* notation because the contribution of a digit to the value of the number depends on where in the expression it sits.

For example, consider the number 0.25. We know this equals $\frac{1}{4}$, but let's belabor the point to be certain the meaning is clear.

- The 0 sits in the ones column (immediately to the left of the decimal point). It contributes $0 \cdot 1$ to the value of the number.

- The 2 sits in the tenths column (immediately to the right of the decimal point). Its contribution is $2 \cdot \frac{1}{10}$.

- Finally, the 5 sits in the hundredths column. It contributes $5 \cdot \frac{1}{100}$ to the value.

Taking it all together, we have this:

$$0.25 = 0 \cdot 1 + 2 \cdot \frac{1}{10} + 5 \cdot \frac{1}{100} = \frac{0}{1} + \frac{2}{10} + \frac{5}{100}$$

$$= \frac{0}{100} + \frac{20}{100} + \frac{5}{100} = \frac{25}{100} = \frac{1}{4}.$$

Unfortunately, there are some numbers that cannot be written in decimal notation if we only use finitely many digits. For example, can we write $\frac{1}{3}$ as a (finite) decimal? If

$$\frac{1}{3} = 0.d_1 d_2 d_3 \ldots d_n$$

where the d_is are single digits, we would have

$$\frac{1}{3} = \frac{d_1}{10} + \frac{d_2}{100} + \frac{d_3}{1000} + \cdots + \frac{d_n}{10^n}.$$

This can be collapsed into the simpler expression with 10^n as the common denominator:

$$\frac{1}{3} = \frac{A}{10^n}$$

where A is a positive integer. Cross multiplying gives this equation:

$$10^n = 3A.$$

Since the right-hand side of this equation is divisible by three, it would follow that 10^n is also divisible by three—which it is not.

In other words, it is impossible to write $\frac{1}{3}$ as a finite decimal.

2.2 What is 0.3333...?

While $\frac{1}{3}$ cannot be written as a finite decimal, let's see that the *infinite* decimal $0.3333\ldots$ is equal to $\frac{1}{3}$. And then we're going to tell you why our analysis is wrong!

What does $0.3333\ldots$ mean? It is an infinite sum:

$$X = \frac{3}{10} + \frac{3}{100} + \frac{3}{1000} + \frac{3}{10000} + \cdots. \tag{2.1}$$

To find the value of X, first multiply both sides by 10:

$$10X = 10 \cdot \left[\frac{3}{10} + \frac{3}{100} + \frac{3}{1000} + \frac{3}{10000} + \cdots \right]$$

$$= 3 + \frac{3}{10} + \frac{3}{100} + \frac{3}{1000} + \frac{3}{10000} + \cdots.$$

From this expression, subtract the expression for X and notice all the cancellations:

$$10X = 3 + \frac{3}{10} + \frac{3}{100} + \frac{3}{1000} + \frac{3}{10000} + \cdots$$

$$- \quad X = \phantom{3 + {}}\frac{3}{10} + \frac{3}{100} + \frac{3}{1000} + \frac{3}{10000} + \cdots$$

$$\therefore \; 9X = 3$$

which gives $X = \frac{1}{3}$.

Looks good? Wrong! Let's try something different using exactly the same analysis. Consider this similar-looking infinite sum:

$$Y = 3 + 3 \cdot 10 + 3 \cdot 100 + 3 \cdot 1000 + 3 \cdot 10000 + \cdots. \qquad (2.2)$$

What is the value of Y? Using the same idea, we start by calculating $10Y$:

$$10Y = 10 \cdot \left[3 + 3 \cdot 10 + 3 \cdot 100 + 3 \cdot 1000 + 3 \cdot 10000 + \cdots \right]$$
$$= 3 \cdot 10 + 3 \cdot 100 + 3 \cdot 1000 + 3 \cdot 10000 + \cdots.$$

Subtract $10Y$ from Y to give:

$$Y = 3 + 3 \cdot 10 + 3 \cdot 100 + 3 \cdot 1000 + 3 \cdot 10000 + \cdots$$
$$- \quad 10Y = \phantom{3 + {}} 3 \cdot 10 + 3 \cdot 100 + 3 \cdot 1000 + 3 \cdot 10000 + \cdots$$
$$-9Y = 3$$

which gives $Y = -\frac{1}{3}$. This is clearly nonsense! The number Y is given as the sum of positive terms. Surely the result can't be negative.

Why do our calculations that determine X from (2.1) give a correct result while the analogous series of calculations give a nonsensical result for Y from (2.2)?

2.3 Infinite sums as sequences

We expressed the infinite decimal $X = 0.3333\ldots$ in equation (2.1) as an infinite sum. We understand what it means to add a finite list of numbers, but how do we understand sums with infinitely many terms? Our experience with Y from equation (2.2) shows that it is easy to make mistakes.

We escape this difficulty by seeing $0.3333\ldots$ not as a sum, but as a *list* of numbers. Here is how to write $0.3333\ldots$ as a list:

$$A = \bigl(0.3, 0.33, 0.333, 0.3333, 0.33333, \ldots\bigr). \tag{2.3}$$

Although the list is infinitely long, it does not include any infinite sums; each entry is a finite decimal.

Now we need to convert the list in (2.3) into a number. We think of the terms in the list as better and better approximations of some real number. To understand that, let's work out an exact expression for the terms in the list (2.3).

Written out in full, the nth term of A is

$$a_n = \frac{3}{10^1} + \frac{3}{10^2} + \frac{3}{10^3} + \cdots + \frac{3}{10^n}. \tag{2.4}$$

This is a finite sum, so ordinary arithmetic steps are legitimate. As we have done before, subtract a_n from $10a_n$ and simplify using the *finitely many* cancellations:

$$10a_n = 10 \cdot \left[\frac{3}{10^1} + \frac{3}{10^2} + \frac{3}{10^3} + \cdots + \frac{3}{10^n}\right]$$
$$= 3 + \frac{3}{10^1} + \frac{3}{10^2} + \frac{3}{10^3} + \cdots + \frac{3}{10^{n-1}}$$
$$-\quad a_n = \frac{3}{10^1} + \frac{3}{10^2} + \frac{3}{10^3} + \cdots + \frac{3}{10^{n-1}} + \frac{3}{10^n}$$

$$\therefore\ 9a_n = 3 - \frac{3}{10^n}$$

which we divide by 9 to conclude

$$a_n = \frac{1}{3} - \frac{1}{3 \cdot 10^n}. \tag{2.5}$$

This exact expression allows us to evaluate the difference between a_n and the value $\frac{1}{3}$ that we believe belongs to the decimal $0.3333\ldots$. We have

$$\left|a_n - \frac{1}{3}\right| = \frac{1}{3 \cdot 10^n}. \tag{2.6}$$

We now can answer this question: How well do the terms of A approximate $\frac{1}{3}$?

Do the terms a_n get within less than 1% of the value $\frac{1}{3}$? Since 1% of $\frac{1}{3}$ is 1/300, we need the difference between a_n and $\frac{1}{3}$ to be less than 1/300. From the fourth term on, the difference between a_n and $\frac{1}{3}$ is at most 3/1000. Since 3/1000 is less than 1/300, we see that from the fourth term on, all elements of sequence A are within 1% of $\frac{1}{3}$.

Let's say we want the approximation to be within one-trillionth of a percent. Does the sequence A eventually get that accurate? It's easy to see that from some large value of n onward, the right-hand side of (2.6) is less than one-trillionth of a percent of $\frac{1}{3}$.

Regardless of how accurate an approximation we desire, from some point on, all terms in sequence A give an estimate of $\frac{1}{3}$ that achieves the desired level of accuracy. The technical term for this is that the sequence A *converges* to $\frac{1}{3}$. By viewing 0.3333... as a sequence converging to $\frac{1}{3}$, we avoid dealing with infinite sums.

Returning to the infinite sum in equation (2.2), if we see that infinite sum as a sequence, we have this:

$$B = \left(3, 33, 333, 3333, \ldots\right).$$

In this case, the terms of this sequence are not increasingly accurate approximations of any real number, so the sum in (2.2) does not converge to any real number.

2.4 Smaller terms but infinite sum

The decimal number 0.3333... can be considered as a sequence, where each term is a little bit bigger than the previous:

$$0$$
$$0 + \tfrac{3}{10}$$
$$0 + \tfrac{3}{10} + \tfrac{3}{100}$$
$$0 + \tfrac{3}{10} + \tfrac{3}{100} + \tfrac{3}{1000}$$
$$\vdots$$

The added terms get progressively smaller. This might lead one to think that as long as the additional terms are shrinking, the ultimate sum can't be too large.

However, this is not the case! We illustrate this phenomenon with a sum of terms that are shrinking, and yet the overall sum is infinite.

Here is the sum:

$$H = 1 + \tfrac{1}{2} + \tfrac{1}{3} + \tfrac{1}{4} + \tfrac{1}{5} + \tfrac{1}{6} + \cdots. \tag{2.7}$$

By the time we get to the millionth term of this sum, the added contribution is a mere 0.000001 and subsequent terms contribute even less.[1] Can the sum H from equation (2.7) really be infinite?

To see that this sum is indeed infinite, we group the terms as follows. We begin by combining the first two terms:

$$H = \underbrace{1 + \tfrac{1}{2}} + \left[\tfrac{1}{3} + \tfrac{1}{4} + \tfrac{1}{5} + \tfrac{1}{6} + \cdots\right]$$
$$= \tfrac{3}{2} + \left[\tfrac{1}{3} + \tfrac{1}{4} + \tfrac{1}{5} + \tfrac{1}{6} + \cdots\right].$$

Next we note that $\tfrac{1}{3} + \tfrac{1}{4} > \tfrac{1}{4} + \tfrac{1}{4} = \tfrac{1}{2}$. Combining those terms, we get:

$$H = \tfrac{3}{2} + \underbrace{\tfrac{1}{3} + \tfrac{1}{4}} + \left[\tfrac{1}{5} + \tfrac{1}{6} + \tfrac{1}{7} + \tfrac{1}{8} + \tfrac{1}{9} + \tfrac{1}{10} + \cdots\right]$$
$$> \tfrac{3}{2} + \tfrac{1}{2} + \left[\tfrac{1}{5} + \tfrac{1}{6} + \tfrac{1}{7} + \tfrac{1}{8} + \tfrac{1}{9} + \tfrac{1}{10} + \cdots\right].$$

Next we add the four terms $\tfrac{1}{5}$ through $\tfrac{1}{8}$, noting that the sum is greater than $4 \cdot \tfrac{1}{8} = \tfrac{1}{2}$:

$$H > \tfrac{3}{2} + \tfrac{1}{2} + \underbrace{\tfrac{1}{5} + \tfrac{1}{6} + \tfrac{1}{7} + \tfrac{1}{8}} + \left[\tfrac{1}{9} + \tfrac{1}{10} + \tfrac{1}{11} + \cdots\right]$$
$$> \tfrac{3}{2} + \tfrac{1}{2} + \tfrac{1}{2} + \left[\tfrac{1}{9} + \tfrac{1}{10} + \tfrac{1}{11} + \cdots\right].$$

Continuing, we add the eight terms $\tfrac{1}{9}$ through $\tfrac{1}{16}$, whose sum is greater than $8 \cdot \tfrac{1}{16} = \tfrac{1}{2}$:

$$H > \tfrac{3}{2} + \tfrac{1}{2} + \tfrac{1}{2} + \underbrace{\tfrac{1}{9} + \cdots + \tfrac{1}{16}} + \left[\tfrac{1}{17} + \tfrac{1}{18} + \cdots\right]$$
$$> \tfrac{3}{2} + \tfrac{1}{2} + \tfrac{1}{2} + \tfrac{1}{2} + \left[\tfrac{1}{17} + \tfrac{1}{18} + \cdots\right].$$

Collecting terms $\tfrac{1}{17}$ through $\tfrac{1}{32}$ yields:

$$H > \tfrac{3}{2} + \tfrac{1}{2} + \tfrac{1}{2} + \tfrac{1}{2} + \tfrac{1}{2} + \left[\tfrac{1}{33} + \tfrac{1}{34} + \tfrac{1}{35} + \cdots\right].$$

Onward we go! Collect the next 32 terms, then the next 64 terms, and so forth:

$$H > \tfrac{3}{2} + \tfrac{1}{2} + \tfrac{1}{2} + \tfrac{1}{2} + \tfrac{1}{2} + \tfrac{1}{2} + \tfrac{1}{2} + \tfrac{1}{2} + \cdots = \infty.$$

---

[1] This infinite sum is known as the *harmonic series*.

For your consideration

As a high school student, I learned that $0.999\ldots = 1$. I also pestered my teacher with the question: What if we have infinitely many 9s after the decimal point, and then there's an 8 at the end?

Why is this utter nonsense? (I forgive my younger self.)

But then again, might we consider the following sequence?

$$\bigl(0.8, 0.98, 0.998, 0.9998, 0.99998, \ldots\bigr)$$

Is this a way to vindicate my younger self? What would be the answer to my question?

Chapter 3

Tropical Arithmetic

> A smile is a curve that sets everything straight.
> —Phyllis Diller

In this chapter, just as in Chapter 1, we incorporate ∞ into the real numbers, $\mathbb{R}$. In that case, we tried, as best we could, to extend the usual operations of addition and multiplication in a way that preserves their fundamental properties.

In this chapter, we once again append ∞ to $\mathbb{R}$, but at the same time we modify the definitions of addition and multiplication. Happily, the new-fangled operations satisfy the desired algebraic properties. The modified real number system we present is called *tropical* (see the comments on the following page for the reason for this name).

Polynomial equations, such as $x^2 + 2y^2 = 1$, create lovely curves in the plane; for this particular equation, the curve is an ellipse. When we use the tropical versions of addition and multiplication, the usual curves of analytic geometry are transformed into shapes made entirely of straight lines. We hope that makes you smile!

3.1 New operations: ⊕ and ⊙

The extended real number system, $\overline{\mathbb{R}}$, from Chapter 1, features both a positive and a negative infinity. As much as possible, we extend the usual arithmetic operations to work with ∞ and −∞.

Tropical arithmetic appends only positive infinity to the real numbers, but replaces ordinary addition and multiplication with two new

> **Why "tropical"?**
>
> This number system is called *tropical* in honor of Imre Simon, a mathematician who lived in Brazil and was one of the early developers of this concept. The word is simply an allusion to the warm climate of Brazil.
>
> Because the two operations for $\mathbb{T}$ are min ($x \oplus y$ is the minimum of x and y) and sum ($x \odot y$ is the sum of x and y), tropical arithmetic is also known as min-sum arithmetic.

operations that behave in ways that are somewhat similar to addition and multiplication.

The numbers in tropical arithmetic are all the real numbers as well as ∞. We use the letter $\mathbb{T}$ to stand for the tropical number system: $\mathbb{T} = \mathbb{R} \cup \{\infty\}$.

The usual operations of addition and multiplication are replaced by *tropical* addition, $\oplus$, and multiplication, $\odot$. The definitions are simple. For tropical numbers x and y, we define

$$x \oplus y = \min\{x, y\}$$
and $$x \odot y = x + y.$$

In words, the tropical sum of x and y is the smaller of the two numbers. For example, $3 \oplus 5 = 3$. The tropical product of x and y is their ordinary sum. For example, $3 \odot 5 = 8$.

These operations may combine real numbers with ∞. We have:

$$x \oplus \infty = \min\{x, \infty\} = x$$
and $$x \odot \infty = x + \infty = \infty.$$

There are no undefined results in tropical arithmetic.

Properties

Many of the algebraic properties enjoyed by ordinary addition and multiplication are also satisfied by their tropical cousins.

Let's begin with comparing $x \oplus y$ and $y \oplus x$. In either case, the result is the smaller of x or y; hence, they are equal. Likewise, $x \odot y$ and $y \odot x$ are both equal to the usual sum $x + y$, so they are equal.

> **Commutative Property for Tropical Addition and Multiplication**
> For any two tropical numbers x and y, we have $x \oplus y = y \oplus x$ and $x \odot y = y \odot x$.

Now let's think about triples of tropical numbers. Notice that

$$x \oplus (y \oplus z) = \min\{x, \min\{y, z\}\} = \min\{x, y, z\}.$$

The same holds for $(x \oplus y) \oplus z$. In both cases, the result is simply the smallest of the three numbers x, y, and z.

For tropical multiplication, the analysis is even simpler:

$$x \odot (y \odot z) = x + (y + z) = (x + y) + z = (x \odot y) \odot z.$$

> **Associative Property for Tropical Addition and Multiplication**
> For any three tropical numbers x, y, and z, we have $x \oplus (y \oplus z) = (x \oplus y) \oplus z$ and $x \odot (y \odot z) = (x \odot y) \odot z$.

The distributive property shows how addition and multiplication interoperate. Let's do the following calculations for tropical numbers x, y, and z.

To begin, we calculate $x \odot (y \oplus z)$. The result of the $\oplus$ term is the smaller of y or z. We then do the $\odot$ operation on that result; that's simply adding x to the result. In other words, $x \odot (y \oplus z)$ is the smaller of $x + y$ or $x + z$.

Now we convert $x + y$ and $x + z$ to tropical language; they are $x \odot y$ and $x \odot z$, respectively. To determine which is smaller, we take their tropical sum: $(x \odot y) \oplus (x \odot z)$.

In other words, both $x \odot (y \oplus z)$ and $(x \odot y) \oplus (x \odot z)$ evaluate to the smaller of $x + y$ and $x + z$. Summarizing:

> **Distributive Property for Tropical Arithmetic**
> For any three tropical numbers x, y, and z, we have $x \odot (y \oplus z) = (x \odot y) \oplus (x \odot z)$.

The identity elements for ordinary addition and multiplication are 0 and 1, respectively. What are they for their tropical cousins?

Recall that $x \oplus y$ is the smaller of x and y. A value for y that would guarantee $x \oplus y = x$ would be $y = \infty$. Since tropical multiplication $\odot$ is actually addition, its identity element is 0.

> **Identity Elements for Tropical Addition and Multiplication**
> Let x be any tropical number. Then $x \oplus \infty = x$ and $x \odot 0 = x$.

Finally, we consider additive and multiplicative inverses. Recall that when additive inverses are added to each other, the result is the identity element for addition. Ordinarily, that would be 0, but we're in a tropical setting. Likewise, multiplying numbers that are multiplicative inverses of each other gives the identity element for multiplication. Ordinarily that would be 1, but not in the tropics.

With this in mind, the additive inverse of x would be another tropical number y so that $x \oplus y = \infty$. However, since $x \oplus y$ is the smaller of x or y, unless $x = \infty$, there is no number y such that $x \oplus y = \infty$; the only way to have $x \oplus y = \infty$ is when $x = y = \infty$. If x is finite, then it does not have an additive inverse.

Most tropical numbers have multiplicative inverses. If $x \neq \infty$, then $x \odot (-x) = x + (-x) = 0$. The only tropical number without a multiplicative inverse is ∞.

This is just like ordinary multiplication in which the only number that does not have a multiplicative inverse is 0, the identity element for addition. Here, the only number that doesn't have a tropical multiplicative inverse is ∞, the identity element for tropical addition.

> **Tropical Multiplicative Inverses**
> If x is a tropical number with $x \neq \infty$, then x has a multiplicative inverse, namely $-x$.

Note that 0 has a multiplicative inverse: itself. This leads to the strange statement that in $\mathbb{T}$ it is possible to divide by 0, but not by ∞.

3.2 Polynomials

Polynomials are expressions that involve one or more variables combined with numbers via the usual addition and multiplication operations. For example, $4x^3 - 5xy + y$ is a polynomial in two variables. Note that x^3 is an alternative way to write $x \cdot x \cdot x$; exponentiation is simply repeated multiplication. Subtraction is not actually needed in writing polynomials because we can rewrite this expression as $4x^3 + (-5)xy + y$.

In tropical arithmetic, we can also add and multiply by replacing ordinary $+$ and $\cdot$ with $\oplus$ and $\odot$. This enables us to create tropical polynomials. To be crystal clear that exponentiation is repeated use of $\odot$, we write $x^{\odot 3}$ instead of x^3. Converting the example polynomial $4x^3 - 5xy + y$ to the tropical context yields this somewhat unsightly expression:[1]

$$4 \odot x^{\odot 3} \oplus (-5) \odot x \odot y \oplus y.$$

Here's a fun fact about tropical polynomials: For ordinary polynomials, the binomial theorem tells us how to expand $(x+y)^n$. For example, $(x+y)^4 = x^4 + 4x^3y + 6x^2y^2 + 4xy^3 + y^4$. The coefficients come from Pascal's triangle. But the analogous calculation in tropical arithmetic is easier. What is $(x \oplus y)^{\odot 4}$? If $x \le y$, the result is $x^{\odot 4} = 4x$. If $y \le x$, the result is $y^{\odot 4} = 4y$. That is, the result of $(x \oplus y)^{\odot 4}$ is the smaller of $x^{\odot 4}$ or $y^{\odot 4}$. That can be neatly written as

$$(x \oplus y)^{\odot 4} = x^{\odot 4} \oplus y^{\odot 4}.$$

In general, if n is a positive integer, we have this remarkably simple identity: $(x \oplus y)^{\odot n} = x^{\odot n} \oplus y^{\odot n}$.

One of the features that makes tropical polynomials interesting is that their graphs are not curved. Consider the simple polynomial $p(x) = x^{\odot 2} \oplus 1$. Let's convert this to ordinary arithmetic. The $x^{\odot 2}$ term, which is just $x \odot x$, becomes $x + x = 2x$:

$$p(x) = x^{\odot 2} \oplus 1 = (2x) \oplus 1.$$

We can further simplify this because $\oplus$ is just the minimum of its arguments:

$$p(x) = x^{\odot 2} \oplus 1 = (2x) \oplus 1 = \min\{2x, 1\}.$$

Note that if $x \le \frac{1}{2}$, then $\min\{2x, 1\}$ evaluates to $2x$. However, for $x \ge \frac{1}{2}$, we see that $p(x)$ evaluates to 1:

$$p(x) = \begin{cases} 2x & \text{if } x \le \frac{1}{2} \\ 1 & \text{if } x \ge \frac{1}{2} \end{cases}$$

---

[1]Some authors enclose tropical expressions inside quotation marks to signify that the operations are $\oplus$ and $\odot$, and not ordinary arithmetic. The example polynomial could also be written as "$4x^3 + (-5)xy + y$".

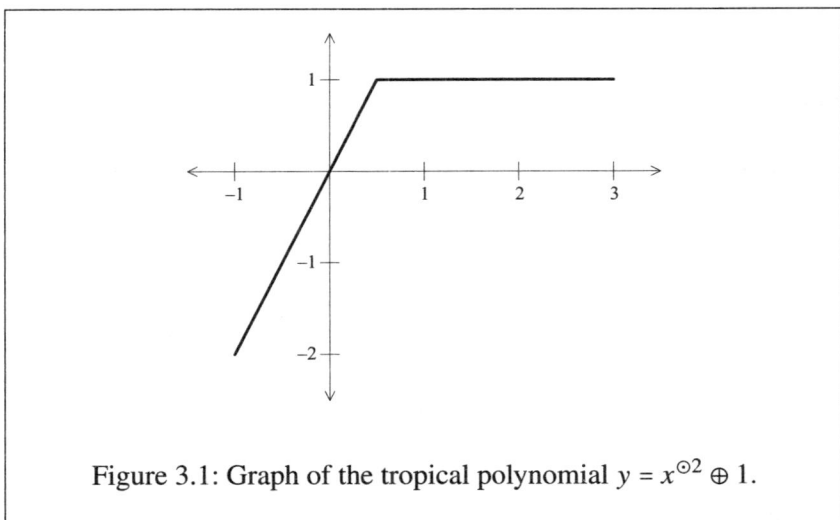

Figure 3.1: Graph of the tropical polynomial $y = x^{\odot 2} \oplus 1$.

We present the graph of $p(x)$ in Figure 3.1. Notice that the graph is composed of two straight pieces: a first part with upward slope connected to a horizontal line.

3.3 Tropical algebraic curves

Curves such as ellipses and parabolas are known as *algebraic curves*. A polynomial in two variables, $p(x, y)$, determines such a curve by considering all points (x, y) for which $p(x, y) = 0$.

For example, consider the polynomial $p(x, y) = x^2 + y^2 - 1$. The set of points (x, y) for which $p(x, y) = 0$ is a circle of radius one centered at the origin.

Another example: Let $p(x, y) = ax + by + c$ (for constants a, b, and c, and forbidding $a = b = 0$). The algebraic curve $p(x, y) = 0$ is a straight line.

What happens when ordinary addition and multiplication are replaced by their tropical counterparts? Although tropical curves look vastly different from their standard cousins, some interesting properties hold in both domains. For example, we know that two lines can intersect in at most one point. We'll see that the same is true in the tropical setting.

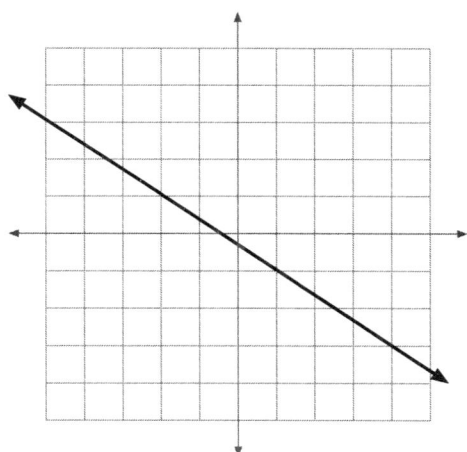

Figure 3.2: For the ordinary polynomial $p(x,y) = 2x + 3y + 1$, the set of points for which $p(x,y) = 0$ is the straight line illustrated here.

Tropical lines

Every line in the plane is the set of points (x,y) for which $p(x,y) = ax+by+c$ is equal to zero. For example, suppose $p(x,y) = 2x+3y+1$. The line $p(x,y) = 0$ includes the points $(-2, 1)$ and $(1, -1)$; it is illustrated in Figure 3.2.

The change to the tropical venue requires two steps. The first is to replace addition and multiplication by their tropical counterparts.

Recasting the polynomial $p(x,y) = 2x + 3y + 1$ using the tropical operations gives this new version:

$$p(x,y) = (2 \odot x) \oplus (3 \odot y) \oplus 1.$$

Let's rewrite this by replacing $\oplus$ with min and $\odot$ with +:

$$p(x,y) = \min\{x + 2, y + 3, 1\}.$$

For example, $p(-1, 0) = \min\{-1 + 2, 0 + 3, 1\} = \min\{1, 3, 1\} = 1$.

The next step is *not* to set $p(x,y) = 0$. Rather, the way mathematicians visualize a tropical polynomial is to consider those points (x,y) for which the minimum value is achieved two (or more) times.

We see that $(-1, 0)$ is such a point. When $x = -1$ and $y = 0$ are substituted into $p(x, y)$, we get the values 1, 3, and 1 for the three terms $2 \odot x$, $3 \odot y$, and 1, respectively. Since the minimum value is repeated, we declare that $(-1, 0)$ is on the curve.

The point $(-2, -3)$ is also on the curve determined by $p(x, y)$. To see this, we evaluate p:

$$p(-2, -3) = \min\{-2 + 2, -3 + 3, 1\} = \min\{0, 0, 1\} = 0.$$

Since the minimum value is repeated, $(-2, -3)$ is on the curve.

However, the point $(1, 2)$ is not on the curve determined by p:

$$p(1, 2) = \min\{1 + 2, 2 + 3, 1\} = \min\{3, 5, 1\} = 1.$$

The minimum value, 1, is not a repeated value.

Let's figure out what this curve looks like. Since $p(x, y) = \min\{x + 2, y+3, 1\}$, to get repeated values, we need one of the following to happen:

(a) $x + 2 = y + 3 \leq 1$, or

(b) $x + 2 = 1 \leq y + 3$, or

(c) $y + 3 = 1 \leq x + 2$.

Each of the equations in (a), (b), and (c) determines a straight line, but the less-than-or-equal condition implies that only half of the line is valid; that is, each of these determines a ray. Notice that the point $(-1, -2)$ satisfies all three of (a), (b), and (c). The result is the Y-shaped plot shown in Figure 3.3.

The intersection of two tropical lines (unless they have overlapping rays) is just a single point. This is an analogue to the situation for ordinary lines that intersect in (at most) one point as in Figure 3.4.

Higher-order curves

We now consider curves arising from polynomials of higher degree.

Polynomials of the form $ax + by + c$ have degree one because the highest exponent in any term is 1. A term of the form $x^s y^t$ is said to have degree $s + t$, and the degree of a polynomial is the highest degree of any of its terms.

Degree-two polynomials have the form

$$p(x, y) = ax^2 + bxy + cy^2 + dx + ey + f$$

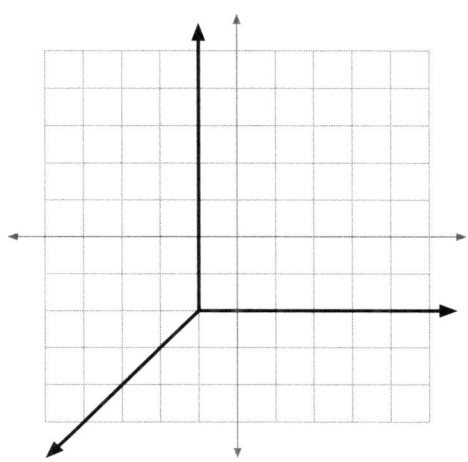

Figure 3.3: Graph of the tropical polynomial
$$p(x,y) = 2 \odot x \oplus 3 \odot y \oplus 1.$$

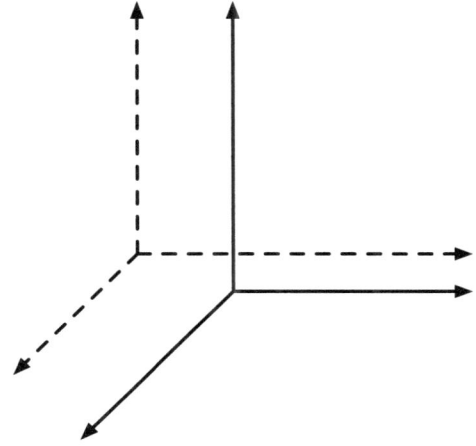

Figure 3.4: Two tropical lines intersecting in a single point.

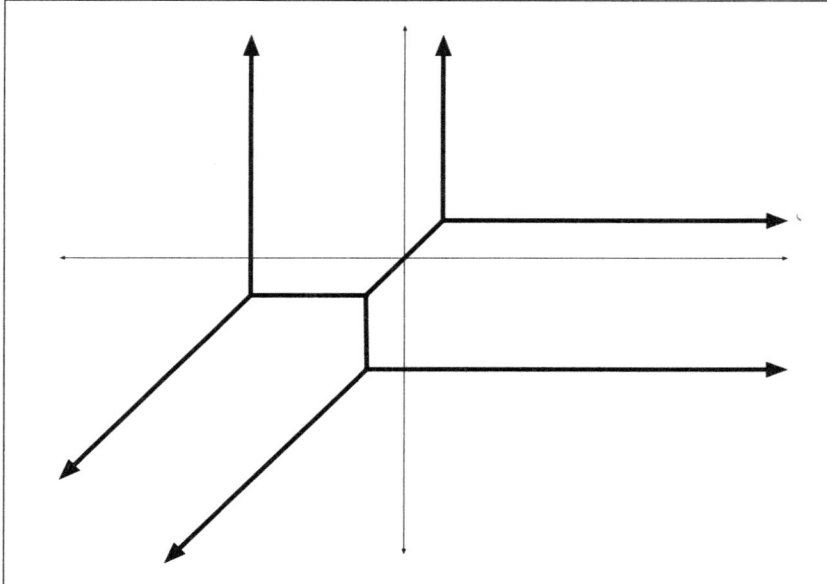

Figure 3.5: Graph of the tropical polynomial

$$p(x,y) = (4 \odot x \odot x) \oplus (1 \odot x \odot y) \oplus (3 \odot y \odot y) \oplus (0 \odot x) \oplus (1 \odot y) \oplus 4.$$

and the (ordinary) curves $p(x,y) = 0$ are the conic sections: ellipses, hyperbolas, and parabolas (and some degenerate versions thereof).

Converting to tropical arithmetic, the function $p(x,y)$ transforms to this:

$$p(x,y) = (a \odot x \odot x) \oplus (b \odot x \odot y) \oplus (c \odot y \odot y) \oplus (d \odot x) \oplus (e \odot y) \oplus f$$
$$= \min\{a + 2x, b + x + y, c + 2y, d + x, e + y, f\}.$$

As before, when we graph the tropical polynomial, we only plot those points (x, y) for which two (or more) of the terms attain the minimum.

Notice that all the terms in the expression for $p(x, y)$ are linear, so the plot will consist entirely of straight pieces: rays and line segments. An example is shown in Figure 3.5.

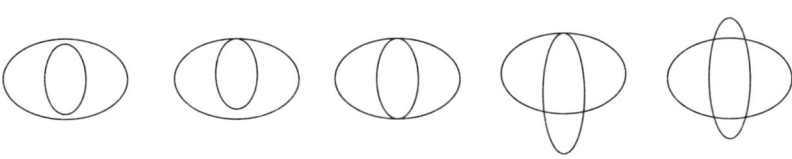

Figure 3.6: Ellipses are graphs of degree-two polynomials. By Bézout's theorem, a pair of ellipses may intersect in at most four points. This figure shows that the number of intersections may be 0, 1, 2, 3, or 4.

Bézout's theorem

The graphs of ordinary and tropical algebraic curves are interesting to mathematicians because they can share common features. A prime example is Bézout's theorem, which counts the number of intersections of two algebraic curves. Bézout's theorem asserts that if $p(x, y)$ is a polynomial of degree d_1 and $q(x, y)$ is a polynomial of degree d_2, then the curves intersect in (at most) $d_1 \cdot d_2$ points. There are some caveats. First, the two curves cannot overlap in infinitely many points; they must only have a finite number of intersecting points. Second, we placed the words "at most" in parentheses because later (in Chapter 6) we show how that caveat can be removed.

Ellipses are algebraic curves of degree two. Therefore, the number of intersections of two ellipses is at most four. In Figure 3.6 we see that two ellipses might intersect in 0, 1, 2, 3, or 4 points, but Bézout's theorem asserts that they cannot intersect in five or more.

Bézout's theorem not only applies to ordinary algebraic curves, but to tropical algebraic curves as well. The graph of a pair of degree-two tropical polynomials is shown in Figure 3.7, where we observe that they intersect in exactly four points.

For your consideration

The tropical number system, $\mathbb{T}$, appends ∞ to the real numbers. How would you incorporate $-\infty$ into $\mathbb{T}$, and how would $-\infty$ work with the $\oplus$ and $\odot$ operations?

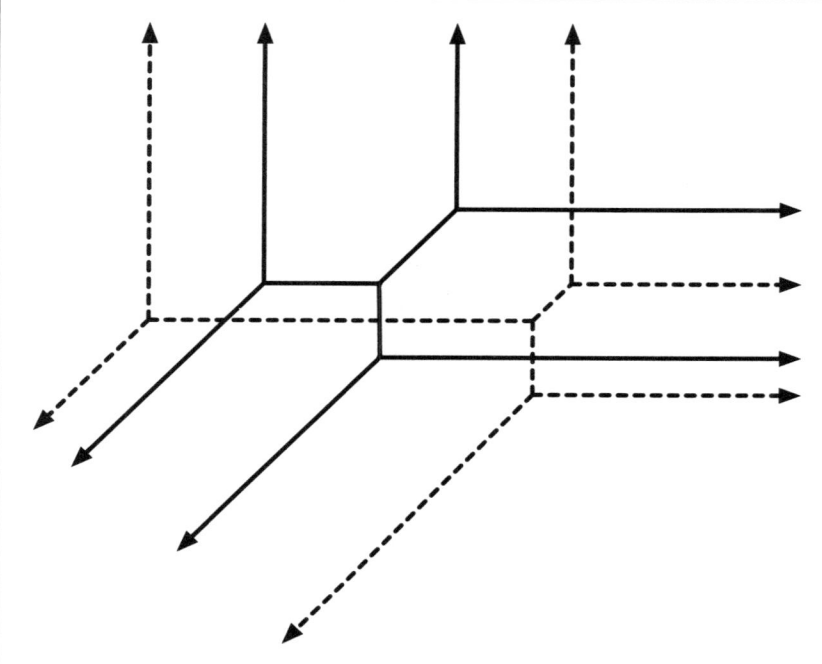

Figure 3.7: Graph of a pair of degree-two tropical polynomials. Notice that the number of points of intersection between these curves is 4, as asserted by Bézout's theorem.

Why is $\ominus$ not an operation for $\mathbb{T}$?

Consider the (ordinary) graphs of $p(x, y) = (x^2 + y^2 - 1)(x - y^2) = 0$ and $q(x, y) = (x^2 + y^2 - 1)(x^2 - y) = 0$. How many points of intersection do they have?

Finally, $x \odot x \oplus y \odot y$ does not equal $x \odot x \oplus y \odot y \oplus 0$. How can you modify one term in the second expression so that it is equal to the first?

Chapter 4

Hyperreals

> Some days Blue wonders why anyone ever bothered making numbers so small; other days she supposes even infinity needs to start somewhere.
>
> —Amal El-Mohtar and Max Gladstone,
> *This Is How You Lose the Time War*

We started our exploration of infinity with a bang: In Chapter 1 we created two incredibly large numbers, ∞ and $-\infty$, and incorporated them with the usual real numbers. This works reasonably well, but certain operations, such as $\infty - \infty$, are left undefined.

In this chapter, we take an entirely different approach altogether. We start with a whimper by introducing a number that is incredibly small: positive, but smaller than any real number. Why do we bother? We use this infinitesimal to create a world of infinities. We had to start somewhere.

4.1 Starting small

The hyperreal numbers, $^*\mathbb{R}$, are a massive extension of the ordinary real numbers. This enormous expansion starts very small: We include infinitesimals.

The hyperreal numbers, $^*\mathbb{R}$, may be thought of as starting with a new positive number, s, that is less than all positive real numbers. Note that s is not a real number, but a new *hyperreal* number.[1] This is illustrated in Figure 4.1.

---

[1] This is a casual description of the hyperreal numbers. The fully rigorous definition—

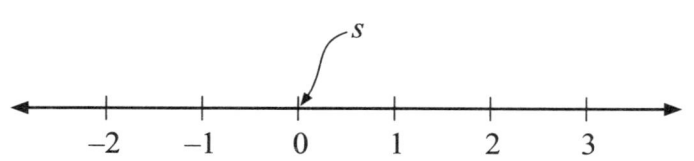

Figure 4.1: The hyperreal numbers, *$\mathbb{R}$, include a positive number, s, that is less than all real numbers. Note that s is not a real number; it is a hyperreal number.

Consecutive real numbers??

In some older books, authors refer to "consecutive" values or points, such as in this excerpt from the textbook *General Geometry and Calculus* by Edward Olney, published in 1871.

44. *A Differential* is the difference between two consecutive states of a function, or variable. It is the same as an infinitesimal.

45. *Consecutive Values* of a function or variable are values which differ from each other by *less than any assignable quantity*.

Consecutive Points on a line are points nearer to each other than any assignable distance.

This is not a meaningful concept. It would be nearly a century before Robinson's theory of hyperreal numbers would give a fully rigorous treatment of infinitesimals.

Having introduced s into the mix, we now allow all the usual arithmetic operations involving s and real numbers. For example, $5 + s - s^2$ is also a hyperreal number, as is $-5/(s+2)$. One implication of this is that between 0 and the positive real numbers, we have infinitely many infinitesimals such as $2s$, $3s$, $4s$, and s^2, s^3, s^4, and many, many more.

---

given by Abraham Robinson—is highly technical. That said, a careful definition of *real numbers* is already daunting.

Not only do we extend the usual arithmetic operations, but we also extend the notion of less-than. We know that $0 < s < x$ where x is any positive real number. In particular, since $s < 1$, by the usual rules of arithmetic, $s^2 < s$. Since s is positive, $2s > s$. Further, $1 + s > 1$, but not by much!

Next we create huge numbers by dividing by s. Consider $1/s$. By the usual rules of $<$, we reason as follows. If x is a positive real number, so is $1/x$. Since s is positive, but smaller than any real number, we know that $s < 1/x$. Inverting, we see that $1/s > x$. In words, $1/s$ is larger than any positive real number—it's infinite!

What's more, $1/s$ is not the only huge number in $^*\mathbb{R}$. The hyperreals also include the numbers $2/s$, $3/s$, and $4/s$, which are all larger still.

Even more dramatically, consider the hyperreal number $1/s^2$. We know that $0 < s^2 < s$ because $0 < s < 1$. Furthermore, s is less than any positive real number; this implies that $s^2 < xs$ where x is any positive real number. This, in turn, implies that $1/s^2 > 1/(xs)$ and so $1/s^2$ is larger than any of $1/s$, $2/s$, $3/s$, and so on. Indeed, it's infinitely larger than $1/s$.

We can keep going: $1/s^3$ is larger than any x/s^2 (where x is a positive real number), and $1/s^4$ is larger than any x/s^3, and so forth.

In $^*\mathbb{R}$ there are infinitely many infinities and infinitesimals! Every finite hyperreal number can be dissected into two parts: a real number plus an infinitesimal.[2] That is, if x is a finite hyperreal number, we can write $x = a + \varepsilon$ where a is an ordinary real number and ε is an infinitesimal (or zero if x is an ordinary real number). The number a is called the *standard part* of x, and we use the notation $\text{st}(x)$ to give the standard part of a. For example, if $x = \frac{5}{8} + s^2$, then $\text{st}(x) = \frac{5}{8}$.

4.2 The transfer principle

The properties of the real numbers are familiar and are reviewed in detail in Section 1.1. In a nutshell, the real numbers, $\mathbb{R}$, are endowed with two operations (addition and multiplication) and an ordering relation (less-than) that obey the following:

- Commutative property for addition and multiplication.

- Associative property for addition and multiplication.

---

[2] A hyperreal number is considered *finite* if it lies between two integers.

- Identity elements for addition (0) and multiplication (1).

- Inverses for addition (negation) and, for nonzero numbers, multiplication (reciprocal).

- Distributive property.

- Harmony between arithmetic and order, such as that the sum and product of positive numbers is positive.

Taken together, we say that $\mathbb{R}$ is an *ordered field*.

In Chapter 1 we needed to sacrifice some of these properties when we extended $\mathbb{R}$ to $\overline{\mathbb{R}}$. For example, ∞ has neither an additive nor a multiplicative inverse. (Recall that $\infty - \infty$ is not 0 and $0 \cdot \infty$ is not 1; both are undefined.)

The situation for $^*\mathbb{R}$ is much better. Many of the properties of the ordinary real numbers $\mathbb{R}$ transfer over to $^*\mathbb{R}$. There is a precise method to translate between ordinary reals and hyperreals, known as the *transfer principle*. We begin by describing the kinds of statements that can be transferred between the two systems.

First-order statements for real numbers

To understand what properties are common to (ordinary) real and hyperreal numbers, we need to explain the kinds of statements to which the transfer principle applies. Let's start with some simple examples.

The commutative property for addition can be written like this:

$$\forall x, \forall y, x + y = y + x. \tag{4.1}$$

The inverted A is a notation that means "for all," "for any," or "for every."[3]

In somewhat stilted language, (4.1) says this:

For all numbers x and for all numbers y, we have $x + y = y + x$.

Here's another example:

$$\forall x, \forall y, \forall z, (x < y \land y < z) \Rightarrow x < z. \tag{4.2}$$

---
[3]The symbol $\forall$ is called the *universal quantifier*.

This statement is the transitive property of the less-than relation. Let's dissect it.

Statement (4.2) begins thus: "For all x, for all y, and for all z." More comfortably, we could say: "For any three numbers x, y, and z."

The next part translates to "if $x < y$ and $y < z$, then $x < z$." The wedge $\wedge$ is a symbol for the logical connective *and*. The double right arrow $\Rightarrow$ is a symbol for an if-then clause.

Putting this all together gives the following verbal translation of (4.2):

For any three numbers x, y, and z, if $x < y$ and $y < z$, then $x < z$.

A further example:

$$\forall x, \forall y, (x < y) \vee (x = y) \vee (x > y). \tag{4.3}$$

The $\vee$ symbol stands for "or." Statement (4.3) can be read like this:

For any numbers x and y, we have $x < y$, or $x = y$, or $x > y$.

Next we show how to express the fact that every real number has an additive inverse. In symbols:

$$\forall x, \exists y, x + y = 0. \tag{4.4}$$

This expression contains a new symbol: the backwards E means "there is" or "there exists."[4] In words, (4.4) reads:

For any number x, there is a number y such that $x + y = 0$.

Expressing the existence of multiplicative inverses can be done like this:

$$\forall x, x \neq 0 \Rightarrow \exists y, x \cdot y = 1. \tag{4.5}$$

In words:

For any number x, if x is not zero, then there is a number y such that $x \cdot y = 1$.

---

[4] The symbol $\exists$ is called the *existential quantifier*.

Sentences that are composed of quantifiers ($\forall$ and $\exists$), arithmetic operations (addition and multiplication), number relations (equality, inequality, less-than), specific numbers (such as 0 or 1), and logical connectives (such as $\wedge$, $\vee$, $\Rightarrow$) are known as *first-order statements*. The term *first-order* refers to the fact that the quantifiers apply to numbers and not to other more complicated structures, such as sets of numbers.

For example, the following sentence is not first order: *Every nonempty set of positive integers contains a least element*. This is a true statement, but the "Every ..." part is not about individual numbers, but rather about sets of numbers.

We are ready to state the transfer principle.[5]

> **Transfer Principle**
> A first-order statement is a theorem for the real numbers, $\mathbb{R}$, if and only if it is a theorem for the hyperreal numbers, $^*\mathbb{R}$.

This is an example of a *meta-theorem*; that is, it is a theorem about theorems. Any first-order statement that we can prove in one context (ordinary reals or hyperreals) must also be true in the other context.

Let's see how the transfer principle makes the inclusion of infinite (and infinitesimal) possible without sacrificing any algebraic properties (sacrifices that are necessary in the context of $\overline{\mathbb{R}}$).

For example, in $\overline{\mathbb{R}}$ there is no additive inverse for ∞. Are there additive inverses for infinite quantities in $^*\mathbb{R}$? The answer is yes because statement (4.4), which is true for real numbers, must also be true for hyperreal numbers.

In $\overline{\mathbb{R}}$ we "tolerated" division by zero, allowing $1/0 = \infty$. This is awkward at best because $0 \cdot \infty$ is undefined (and not equal to 1).

Can we divide by zero in $^*\mathbb{R}$? To be clear, the expression $a \div b = c$ means that $b \cdot c = a$. Is there a value for $1 \div 0$ in $^*\mathbb{R}$? If so, it would be a number x such that $0 \cdot x = 1$. However, the following is true for real numbers:

$$\forall x, x \cdot 0 = 0. \tag{4.6}$$

Since (4.6) is true for $\mathbb{R}$, by the transfer principle, it is also true in $^*\mathbb{R}$. Hence, there is no legitimate value for $1 \div 0$ in $^*\mathbb{R}$.

---

[5] The proof of the transfer principle is highly technical. We discuss the principle here because it is a cornerstone of the theory of the hyperreal numbers.

The hyperreal numbers $^*\mathbb{R}$ are a massive extension of the ordinary real numbers $\mathbb{R}$ that add infinitesimal values, infinite values, and, for each real number x, infinitely many nonstandard numbers that are infinitely close to, but different from, x.

Though it is hard to imagine, there is a truly surreal[6] number system that extends $^*\mathbb{R}$.

4.3 Derivatives made easy

The *derivative* of a function is a fundamental concept in calculus that rests on the more basic concept of a *limit*.[7] We show here how to calculate derivatives using nonstandard analysis.

The concept of a derivative is based on the notion of *slope*. Given a line in the coordinate plane, choose two distinct points on that line: (x_1, y_1) and (x_2, y_2). The *slope* of the line is

$$\frac{\Delta y}{\Delta x} = \frac{y_2 - y_1}{x_2 - x_1}$$

as illustrated in Figure 4.2.

The notation Δy means "change in y"; likewise for Δx. If the line is horizontal, the slope is 0. If the line proceeds from the upper left to the lower right, the slope is negative. Further, if the line is vertical, then we declare the slope to be infinite. We revisit lines and their slopes in Chapter 6.

Derivatives are slopes. Imagine a function f whose graph is a nice, smooth curve. The derivative of f at the point x is the slope of the curve at the point $(x, f(x))$.

As a concrete example, consider the graph of the function $y = x^2 + 1$ shown in Figure 4.3. Because this graph is curved, there is no single number that describes its slope. For negative x the graph is sloping downward, and for positive x it turns upward.

A point on the curve has coordinates $(x, x^2 + 1)$. We can draw a line through that point that is tangent to the curve and use that to specify the slope at that point. See the dashed line in Figure 4.3.

---

[6] We recommend Donald Knuth's charming book *Surreal Numbers*.
[7] Readers who have not studied calculus should not be deterred! We will see how to find derivatives without taking limits. Readers who have studied calculus will delight in the simplicity that hyperreal numbers afford.

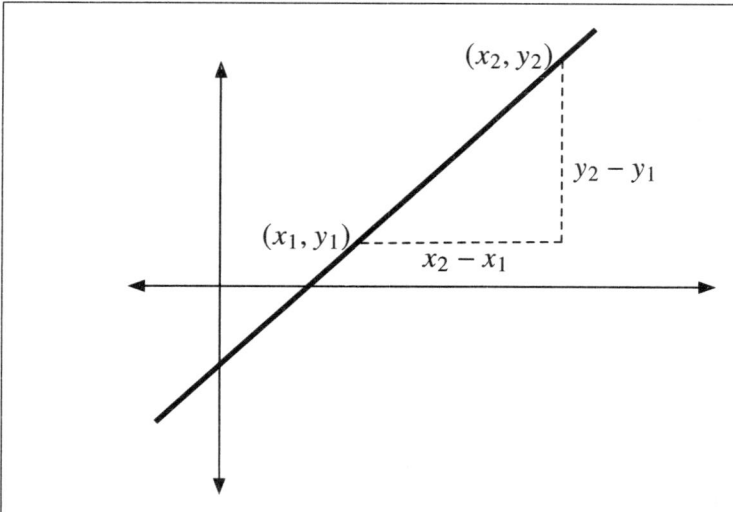

Figure 4.2: The slope of a line is calculated by considering two points on the line: (x_1, y_1) and (x_2, y_2). The slope is $(y_2 - y_1)/(x_2 - x_1)$.

To calculate the slope of the line at a given point on the parabola, we switch from $\mathbb{R}$ to $^*\mathbb{R}$. We calculate the slope of the line joining the point $(x, x^2 + 1)$ to a point infinitesimally close by on the curve: $(x + s, (x + s)^2 + 1)$. We get this:

$$\begin{aligned}\frac{\Delta y}{\Delta x} &= \frac{[(x+s)^2 + 1] - [x^2 + 1]}{(x+s) - x} \\ &= \frac{[x^2 + 2sx + s^2 + 1] - [x^2 + 1]}{s} \\ &= \frac{2sx + s^2}{s} \\ &= 2x + s.\end{aligned}$$

We now convert this result from a hyperreal number to a real number simply by taking the standard part (st) of the answer:

$$\text{slope} = \text{st}\left[\frac{\Delta y}{\Delta x}\right] = \text{st}(2x + s) = 2x.$$

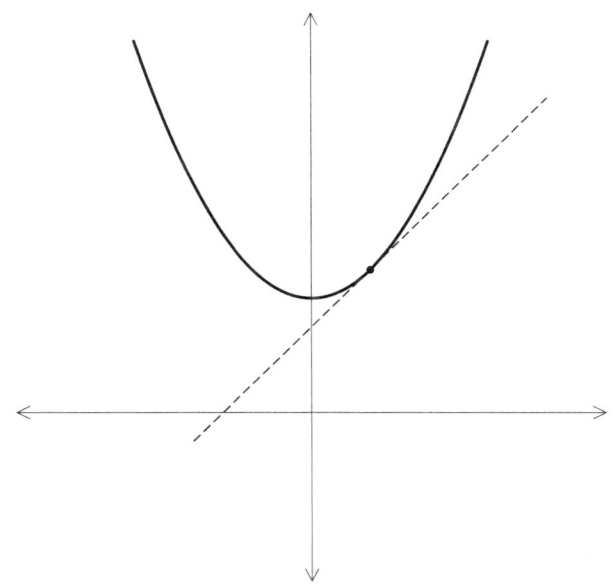

Figure 4.3: Graph of the function $y = x^2 + 1$ and a line tangent to the parabola at the point $(x, x^2 + 1)$. The slope of the tangent line is the derivative of the function $y = x^2 + 1$ at that point. We can calculate the slope of that line with the help of hyperreal numbers.

Using standard notations from calculus, we have that the derivative of $y = x^2 + 1$ at x is:
$$y' = \frac{dy}{dx} = 2x.$$

For your consideration

Let A be the set of all rational numbers whose square is less than 2. In symbols:
$$A = \{x \in \mathbb{Q} : x^2 < 2\}.$$
For example, $\frac{7}{5}$ is an element of A because
$$\left(\frac{7}{5}\right)^2 = \frac{7}{5} \cdot \frac{7}{5} = \frac{49}{25} < 2.$$

However, $\frac{3}{2}$ is not a member of A because

$$\left(\frac{3}{2}\right)^2 = \frac{3}{2} \cdot \frac{3}{2} = \frac{9}{4} > 2.$$

Although $\frac{3}{2}$ is not in A, it is an upper bound for A because all elements of A are no larger than $\frac{3}{2}$. In general, b is an *upper bound* for a set X if $x \le b$ for any number x in the set X.

Of course, $\frac{3}{2}$ is not the only upper bound for A. All elements of A are less than 2, so 2 is also an upper bound for A.

A *least upper bound* for a set X is a number b with these properties:

- b is an upper bound for the set X, and
- if b' is any other upper bound for X, then $b' > b$.

For example, let N be the set of negative real numbers. Then 2 is an upper bound for N, as are 1 and $\frac{1}{2}$. In the context of the real numbers, 0 is the least upper bound for N. However, in the context of $^*\mathbb{R}$, the set N does not have a least upper bound because $-s$ (where s is an infinitesimal) is also an upper bound for N (why?) as are $-2s$, $-3s$, $-4s$, and so forth.

Returning to the set $A = \{x \in \mathbb{Q} : x^2 < 2\}$, we ask if A has a least upper bound in the following contexts:

- Does A have a least upper bound in the context of $\mathbb{Q}$?
- Does A have a least upper bound in the context of $\mathbb{R}$?
- Does A have a least upper bound in the context of $^*\mathbb{R}$?

The real number system, $\mathbb{R}$, enjoys the property that any nonempty subset of $\mathbb{R}$ that has an upper bound must have a least upper bound. This property does not hold for $^*\mathbb{R}$. Why isn't this a violation of the transfer principle?

Part II: Plane Infinity

Chapter 5

Complex Numbers, Extended

> I'm kinda tired. I was up all night trying to round off infinity.
> —Steven Wright

The real numbers, $\mathbb{R}$, are nicely visualized as a number line. In Chapter 1 we introduced two new numbers, ∞ and $-\infty$, to create the extended real numbers, $\overline{\mathbb{R}}$. We place these numbers "all the way" beyond the ends of the line (as in Figure 1.1).

In this chapter we extend the complex numbers, $\mathbb{C}$, to include infinity. Just as the real numbers can be viewed geometrically as a line, the complex numbers are nicely represented as a plane. How do we incorporate ∞ into $\mathbb{C}$ geometrically? We do this by rounding off the complex numbers into a ball. This will not take us all night!

5.1 Complex numbers

In Part I, we considered various ways in which infinity may be incorporated into the real numbers, $\mathbb{R}$. In this chapter, our attention turns to the complex numbers, $\mathbb{C}$. Let's have a quick reminder of what these are.

Complex numbers arise from the real numbers by the inclusion of a new number, i. The number i has the property that $i^2 = -1$; that is, i is a square root of -1. The full set of complex numbers emerges by allowing the imaginary i to be added and multiplied by real numbers.[1] Every

---
[1] The use of the word *imaginary* is unfortunate. The number i is no less "real" than real numbers. All numbers are mental constructs. Real numbers might feel more "real" because they reasonably model measurements.

complex number can be written in the standard form $a + bi$ where a and b are real numbers. The basic operations of addition and multiplication work as follows.

The sum of complex numbers $a + bi$ and $c + di$ is formed simply by adding their respective real parts (a and c) and their respective imaginary parts (b and d). Specifically:

$$(a + bi) + (c + di) = (a + c) + (b + d)i.$$

For example, the sum of $3 + 5i$ and $4 - i$ is $7 + 4i$.

The product of complex numbers follows the usual rules of algebra:

$$(a + bi) \cdot (c + di) = ac + adi + bci + bdi^2$$
$$= (ac + bdi^2) + (ad + bc)i$$
$$= (ac - bd) + (ad + bc)i.$$

For example, the product of $3 + 5i$ and $4 - i$ is $17 + 17i$.

It's easy to check that $0 + 0i = 0$ is the identity element for addition and $1 + 0i = 1$ is the identity element for multiplication.

The additive inverse of $a + bi$ is $-(a + bi) = -a - bi$. Multiplicative inverses are more complicated. To write the complex number $1/(a + bi)$ in the form $X + Yi$, we multiply the numerator and denominator by $a - bi$:

$$\frac{1}{a + bi} = \left[\frac{1}{a + bi}\right] \cdot \left[\frac{a - bi}{a - bi}\right] = \frac{a - bi}{(a + bi)(a - bi)}$$
$$= \frac{a - bi}{a^2 - abi + abi - b^2i^2} = \frac{a - bi}{a^2 + b^2} = \left[\frac{a}{a^2 + b^2}\right] - \left[\frac{b}{a^2 + b^2}\right]i.$$

For example, the multiplicative inverse of $3 + 5i$ is

$$\frac{3}{34} - \frac{5}{34}i.$$

To check that is correct, we multiply:

$$[3 + 5i] \cdot \left[\frac{3}{34} - \frac{5}{34}i\right] = \left[3 \cdot \frac{3}{34} + \frac{25}{34}\right] + \left[\frac{-15}{34} + \frac{15}{34}\right]i$$
$$= \left[\frac{9}{34} + \frac{25}{34}\right] + 0i = 1.$$

In short, all the usual algebraic properties for the basic operations (addition, subtraction, multiplication, and division) work exactly as expected for complex numbers.

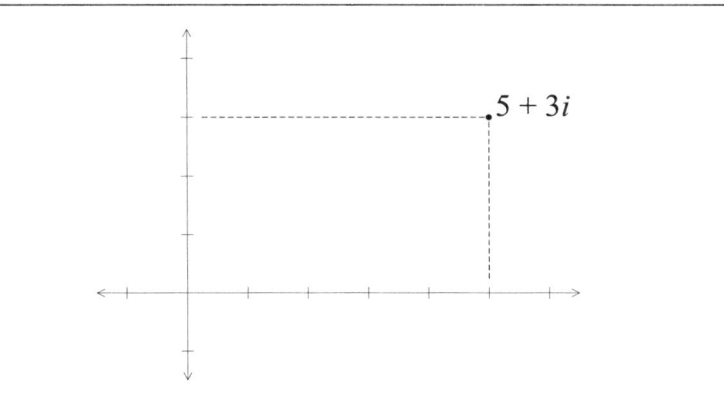

Figure 5.1: The number $5 + 3i$ is represented in the complex plane as a point at location $(5, 3)$.

One feature we have for real numbers but not for complex numbers is a less-than relation. There's no way to define $<$ for $\mathbb{C}$ that is compatible with the usual rules of algebra. This implies that one shouldn't visualize the complex numbers in a linear fashion. Instead, rather than a number line for the reals, we have the *complex plane*.

The complex plane is just the usual plane, but instead of naming points using x- and y-coordinates, we name each point with a single complex number. That is, rather than naming a point (x, y), we name it $x + yi$. This is illustrated in Figure 5.1.

For the real line, it is natural to append ∞ and $-\infty$ "all the way" to the right and left of the number line. What shall we do in the case of complex numbers?

5.2 One infinity for all

The natural way to incorporate infinity into the real numbers is to append ∞ and $-\infty$ at either "end" of the real line (as in Chapter 1). In the case of complex numbers, we are faced with a variety of choices.

For example, one might allow complex numbers $a + bi$ where a and/or b could be either ∞ or $-\infty$. This approach to extending complex numbers is implemented by some computer programming languages such as Julia.

Alternatively, we could create an entire "circumference" of infinities by extending the complex plane. This is akin to the approaches we take in Chapters 6 and 7.

These approaches are reasonable, but the arithmetic becomes exceedingly complicated. Instead, the simplest approach is the one usually adopted for complex numbers: append a single ∞ to the complex numbers and define the extended complex numbers as $\overline{\mathbb{C}} = \mathbb{C} \cup \{\infty\}$.

All that remains is to define how this single, complex ∞ behaves in arithmetic operations.

- Addition: For any finite complex number z, we define $z + \infty = \infty$. However, convention is to leave $\infty + \infty$ undefined.

- Multiplication: For any nonzero complex number z, define $z \cdot \infty = \infty$. We have $\infty \cdot \infty = \infty$ but leave $0 \cdot \infty$ undefined. This implies that $-\infty = -1 \cdot \infty = \infty$; again, there is only one infinity.

- Subtraction: Because $-\infty = \infty$, subtraction behaves just like addition. For finite z, we have $\infty - z = z - \infty = \infty$. As in the case of addition, $\infty - \infty$ is undefined.

- Division: For any finite complex number z, we have $z \div \infty = 0$ and $\infty \div z = \infty$. Provided $z \neq 0$, we have $z \div 0 = \infty$.

  The only divisions that are problematic are $0 \div 0$ and $\infty \div \infty$, which we choose to leave undefined.

5.3 Stereographic projection and the Riemann sphere

The extended real numbers, $\overline{\mathbb{R}}$, can be visualized as a standard number line to which we attach $-\infty$ and ∞ at the far left and far right. Strictly speaking, this doesn't make sense: How far to the left is the "far left"!?

However, in the case of the extended complex numbers, $\overline{\mathbb{C}}$, there is an elegant way to geometrically incorporate ∞ with the (ordinary) complex numbers, $\mathbb{C}$. The idea is to "wrap" the complex plane onto a ball. This is the opposite of the work of a cartographer who needs to represent regions on the globe upon a flat surface.

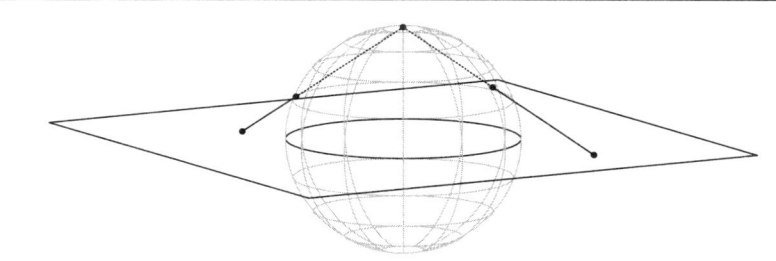

Figure 5.2: A three-dimensional view of stereographic projection. Points on the sphere are projected onto points in the complex plane. Two example projections are shown. The projection from the north pole of the sphere is shown as a line segment through a point on the sphere and onto a point in the complex plane; the portion of that segment that is interior to the sphere is drawn with a dotted line.

Not shown: A point on the southern hemisphere would project to a point in the complex plane that lies in the interior of the equator (shown as a dark circle).

The method we use is called *stereographic projection*. Imagine the complex plane, $\mathbb{C}$, as lying inside three-dimensional space. Place a sphere of radius one centered at the origin.

Points in the plane are paired with points on the sphere. This correspondence is created by drawing a line joining the complex number on the plane to the north pole of the sphere. The intersection of that line with the sphere gives the point on the sphere that corresponds to the complex number. This is illustrated in Figures 5.2 and 5.3.

In this way, every point in the complex plane corresponds to a point on the sphere, and—nearly—vice versa: almost every point on the sphere corresponds to a point in the plane, except one. The exception is the north pole of the sphere, which does not have a projection into the plane. Instead, the north pole corresponds to ∞, and in this way the points on the sphere encompass all elements of $\overline{\mathbb{C}}$.

This representation of the extended complex numbers, $\overline{\mathbb{C}}$, is known as the *Riemann sphere* in honor of the nineteenth-century German mathematician Bernhard Riemann.

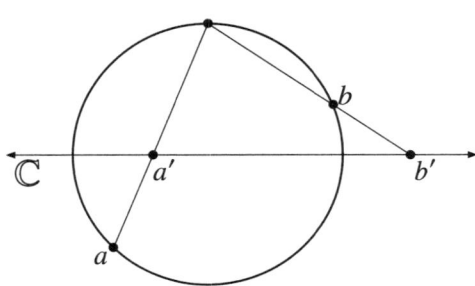

Figure 5.3: A side view of stereographic projection. The horizontal line represents the complex plane, $\mathbb{C}$. Points a and b on the sphere correspond to the numbers a' and b' in the complex plane.

Circles and lines

There is a delightful connection between circles and lines in the complex plane with circles on the Riemann sphere. Slice the Riemann sphere with a plane; that intersection is a circle on the sphere. Now project that circle onto the complex plane, and the result is a circle in the plane—that is, unless the original circle on the sphere runs through the north pole. In that case, the projection is a straight line in the plane.

And this correspondence works the other way. Take any circle or any line in the plane, and project it onto the sphere. The result will be a circle on the sphere. If we began with a line in the plane, then the projection up to the sphere gives a circle that runs through the north pole.

5.4 Linear fractional transformations

In this section we consider a unified approach to three types of functions defined for complex numbers:

- $f(z) = z + a$ where a is a specific complex number,
- $f(z) = az$ where a is a specific, nonzero complex number, and
- $f(z) = 1/z$.

In all three cases, we can extend the functions to include ∞:

- $\infty + a = \infty$,
- $a \cdot \infty = \infty$ (provided $a \neq 0$), and
- $1/\infty = 0$ (as well as $1/0 = \infty$).

These three types of functions are all instances of *linear fractional transformations*. These are defined to be functions of the form

$$f(z) = \frac{az+b}{cz+d}.$$

The one caveat is that the numerator should not be a multiple of the denominator, and that is tantamount to requiring that $ad - bc \neq 0$.

How does ∞ work with these functions?

- If $c \neq 0$, then we define[2] $f(\infty) = a/c$. Further, $f(-d/c) = \infty$.
- If $c = 0$, then necessarily $a \neq 0$ and we have $f(\infty) = \infty$.

Let's see how these functions act geometrically.

The function $f(z) = z + a$ is the easiest to understand. This function shifts the complex plane based on the number a. For example, if $a = 3 + 5i$, then the point z is moved 3 units to the right and 5 units vertically. If we think about this for the extended complex numbers, what we have done is simply shifted the position of the Riemann sphere.

The function $f(z) = az$ (with $a \neq 0$) is more interesting. Suppose $a = x + yi$ where x and y are real numbers. The trick is to use some trigonometry: Rewrite a as

$$a = x + yi = (r\cos\theta) + (r\sin\theta)i = r(\cos\theta + i\sin\theta)$$

where r and θ are real numbers. The pair (r, θ) are known as the *polar coordinates* of the complex number. This is illustrated in Figure 5.4.

---

[2] If z is an enormous complex number, then $(az+b)/(cz+d)$ is very nearly a/c, thereby justifying this choice. This somewhat goes against our earlier decision to leave ∞/∞ undefined.

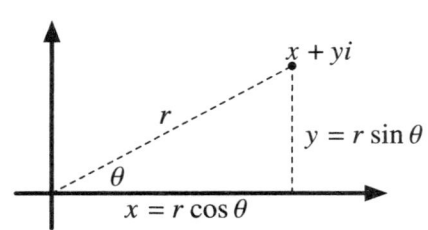

Figure 5.4: The complex number $x + yi$ expressed in polar form as $r(\cos\theta + i\sin\theta)$.

To do this, we need
$$x = r\cos\theta \quad \text{and} \quad y = r\sin\theta.$$
The radius r can be determined from $x^2 + y^2$ because
$$x^2 + y^2 = r^2\cos^2\theta + r^2\sin^2\theta = r^2\left(\cos^2\theta + \sin^2\theta\right) = r^2$$
and so $r = \sqrt{x^2 + y^2}$.

The angle θ can be determined from the fact that
$$\frac{y}{x} = \frac{r\sin\theta}{r\cos\theta} = \frac{\sin\theta}{\cos\theta} = \tan\theta.$$

We need to understand how multiplication behaves geometrically. Write the complex numbers a and b in their polar forms:
$$a = r(\cos\theta + i\sin\theta) \quad \text{and} \quad b = s(\cos\phi + i\sin\phi).$$
Multiply a and b:
$$\begin{aligned} ab &= [r(\cos\theta + i\sin\theta)][s(\cos\phi + i\sin\phi)] \\ &= rs\left[(\cos\theta\cos\phi - \sin\theta\sin\phi) + i(\sin\theta\cos\phi + \cos\theta\sin\phi)\right] \\ &= rs\left[\cos(\theta + \phi) + i\sin(\theta + \phi)\right] \end{aligned}$$
where the final calculation follows from the angle-sum formulas for sine and cosine.

The polar form of ab reveals what has happened geometrically.

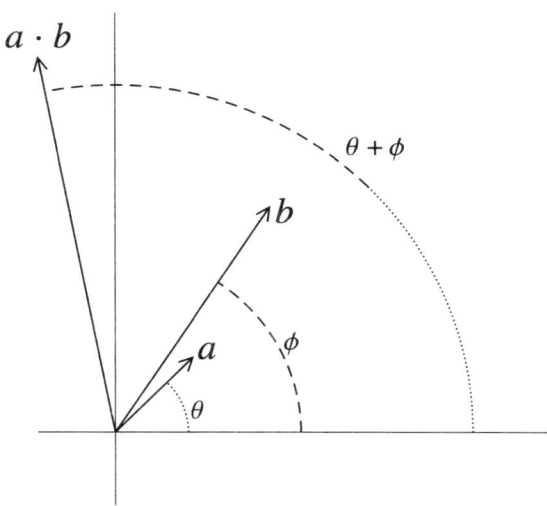

Figure 5.5: An illustration of the multiplication of complex numbers a and b. The radius of $a \cdot b$ is the product of the radii of a and b. The angle of $a \cdot b$ is the sum of the angles of a and b.

- The radius of ab is the result of multiplying the radii of a and b.
- The angle of ab is the result of adding the angles of a and b.

This is illustrated in Figure 5.5.

We can now describe the function $f(z) = az$ in geometric terms. Writing a in the form $r(\cos\theta + i\sin\theta)$, we see that the transformation from z to az rescales z by a factor of r and rotates its angle by θ. In other words, f is a rescaling and a rotation.

From the perspective of the Riemann sphere, we can think of the action of f as rescaling the sphere and rotating the sphere around an axis between its north and south poles.

Finally, let's consider the function $f(z) = 1/z$ by examining its action on the Riemann sphere. First, notice that $f(1) = 1$ and $f(-1) = -1$; the function f does not change those two positions on the sphere. However, $f(i) = 1/i = -i$ and $f(-i) = 1/(-i) = i$; those positions on the sphere are swapped: $i \leftrightarrow -i$. Further, notice that $f(0) = \infty$ and $f(\infty) = 0$; that's a swap of the north and south poles; that is, $0 \leftrightarrow \infty$.

What is happening is that the Riemann sphere has been rotated through 180° along the x-axis (the axis through 1 and -1). The effect of this transformation on a line and a circle is illustrated in Figure 5.6.

Linear fractional transformations are combinations of these basic types of functions:

- $f(z) = z + a$ shifts the Riemann sphere to a new position.

- $f(z) = az$ rescales the Riemann sphere and rotates it along its north-south axis.

- $f(z) = 1/z$ rotates the Riemann sphere 180° degrees through the axis joining 1 and -1.

Combining these ultimately means that linear fractional transformations just shift, roll, and/or magnify (shrink) the Riemann sphere. The north pole, which represents ∞, gets no special treatment. In this way, incorporating ∞ into the complex numbers is quite natural.

For your consideration

The complex numbers, $\mathbb{C}$, arise from the real numbers, $\mathbb{R}$, by the inclusion of the imaginary number i and allowing the usual arithmetic operations. The result is that every complex number has the form $a + bi$ where a and b are real numbers. In notation, $\mathbb{C} = \mathbb{R}[i]$.

What would be the result of appending i to the extended real numbers, $\overline{\mathbb{R}}$ (see Chapter 1)? Let's notate the resulting number system by $\overline{\mathbb{R}}[i]$.

The system $\overline{\mathbb{R}}[i]$ contains all numbers of the form $a + bi$ where a is a real number, is ∞, or is $-\infty$, and likewise for b. How should operations in $\overline{\mathbb{R}}[i]$ act?

For example, if $w = 3 + \infty i$ and $z = \infty + 3i$, what values would you assign to $w + z$, $w - z$, $w \cdot z$, and w/z, or would you leave some of these undefined?

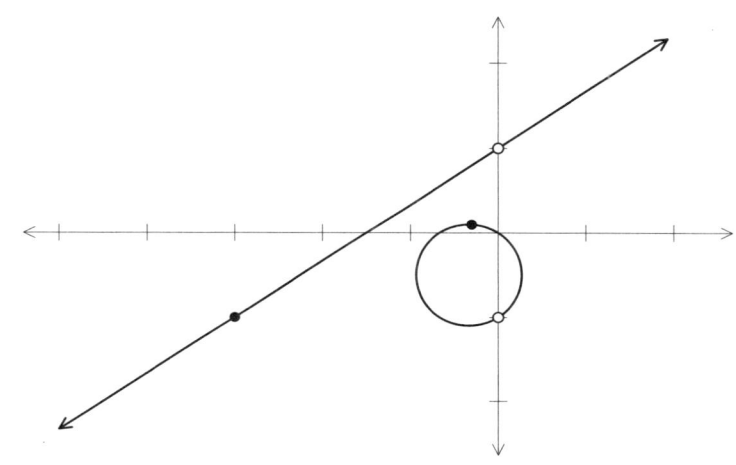

Figure 5.6: The function $f(z) = 1/z$ swaps 0 and ∞; that is, $f(0) = \infty$ and $f(\infty) = 0$. This figure illustrates what happens when f is applied to all the points on a line, L.

The result is a circle. Since L goes to infinity, the circle passes through the origin, $0 + 0i$. Two other points on L are highlighted. The black point $-3 - i$ is on L and its reciprocal $-0.3 + 0.1i$ is on the circle. Likewise, the white point at $0 + i$ is on L and its reciprocal $0 - i$ is on the circle.

Chapter 6

Line at Infinity I: Projective Plane

> ... nice parallel lines make me sick.
> —Eva Hesse, interview in *Artforum*

In this chapter, we enter a geometric universe that has banished parallel lines. Travel with us to infinity on the projective plane: We hope the trip to where parallel lines meet doesn't make you airsick!

6.1 Parallel lines meet at infinity

Look at a long set of railroad tracks heading straight toward the horizon. Rather than appearing equidistant (as they are), it seems that they are coming together and will meet at some distant point. This view of parallel lines—that they give the illusion of meeting way off in the distance—is an artifact of how our eyes receive images from the world. Parallel lines appear to meet "at infinity."

Renaissance artists capitalized on this apparent convergence to create paintings that are much more realistic than works from previous eras. A drawing of a simple box shows how this is done (Figure 6.1). Rather than drawing the sides of the box with parallel lines, a better sense of depth is created by having the sides of the box, when extended, meet at points on an imaginary horizon.

In the painting *A Man Weighing Gold*, one set of parallel lines meets at a point at the left, and a perpendicular set of parallel lines meets at a point

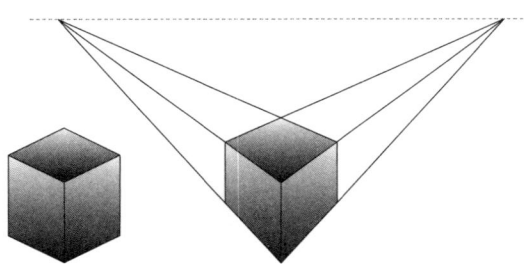

Figure 6.1: Two ways to draw a box. The sides of the box at the left are parallelograms. That is, opposite sides of each face are parallel. The other box is drawn using two-point perspective. Extending one set of edges of the box leads to a point on the horizon (the dashed line). Extending the edges perpendicular to the first set leads to another point on the horizon.

at the right (Figure 6.2). By careful use of this technique of perspective, the seventeenth-century Dutch artist Cornelis de Man creates a highly realistic image.

Of course, parallel lines do not meet—that is, unless we extend them to include a point "at infinity."

6.2 The projective plane

The Euclidean plane—the standard plane of geometry—includes lines that never meet. Parallel lines are equally far apart no matter where we look. However, we can capture the idea that they "eventually" meet by simply giving each of them an extra point.

This idea is similar to what we did in Chapter 1 (in which we added two extra points to the real number line) and in Chapter 5 (in which we added one extra point to the complex plane).

As in those cases, we simply make up some new points: one new point for each set of parallel lines. For example, consider all the horizontal lines in the plane. We manufacture a new point that we add to all of those lines. Or consider all the vertical lines. We make up a different new point, and add it to all the vertical lines.

Figure 6.2: *A Man Weighing Gold* by Cornelis de Man, c. 1670. In this painting, the edges of the tiles in the floor provide two sets of parallel lines. One set includes a seam between tiles, the edge of the table, and a thick beam in the ceiling; these lines meet at a point (outside the frame of the painting) at the left. The lines perpendicular to this first set include a seam between tiles, the mantle of the fireplace, and ceiling rafters. These lines meet at a point on the right.

In other words, for each possible slope we create one new point.[1] All of the lines in the plane with a given slope are augmented with a common new point corresponding to their slope.

Voila! Parallel lines (those with the same slope) all meet at one of these new points. We say that these new points are "at infinity," though we don't actually worry about exactly where that is. Furthermore, we consider all of the points at infinity to be a line as well, which we call *the line at infinity*.

We have extended the ordinary, Euclidean plane to create a new structure called the *projective plane*.

---

[1] Slope is defined in Section 4.3.

An interesting feature of the projective plane is that it establishes a symmetry between points and lines that the Euclidean plane lacks.

In the Euclidean plane, any two distinct points determine a unique line. That is, given two different points, A and B, there is a line—and only one line—that contains both A and B. However, given two distinct lines, there might or might not be a point common to them both (depending on whether they intersect or are parallel).

However, in the projective plane, there is a lovely duality:

- *Given any two distinct points, there is a unique line that contains them both.*

    - The two points might be "ordinary" points, in which case the line is the usual Euclidean line plus its extension point.

    - One point might be "ordinary" and the other point might be "at infinity." In this case, there still is a unique line because given a point and the requisite slope, there is a unique line containing that point with the given slope.

    - Both points might be "at infinity." In this case, the unique line containing the two points is the line at infinity.

- *Given any two lines, there is a unique point contained in both of them.*

    - The two lines might be (extended) "ordinary" lines that are not parallel. The point that they both contain is just the usual point of intersection in the Euclidean plane.

    - The two lines might be (extended) "ordinary" lines that are parallel. In this case, they have the same slope, so they contain the same point at infinity.

    - One line might be an (extended) "ordinary" line, with the other one the line at infinity. In this case, the extended ordinary line contains one point at infinity, and that is the intersection.

Figure 6.3 provides a visualization of how, in a projective plane, any two points (ordinary or at infinity) determine a line and how any two lines (ordinary or at infinity) intersect in a point.

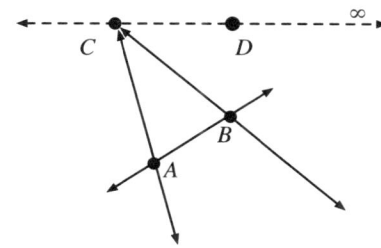

Figure 6.3: The dashed line in this figure represents the line at infinity, and the solid lines represent ordinary lines.

Here, ordinary points A and B determine a unique line, ordinary point B and point at infinity C determine a unique line (from the point B and the slope C), and two points at infinity, C and D, determine a unique line (the line at infinity).

Similarly, a pair of ordinary lines intersect at A, parallel lines AC and BC intersect at infinity (at C), and an ordinary line AC and the line at infinity intersect at C.

6.3 Points and lines, algebraically

Equations for lines

In the (standard) plane, points may be represented by a pair of numbers: (s, t) where s and t are real numbers. These are the coordinates of the point.

Most lines can be represented by an equation of the form $y = mx + b$. The points (x, y) that satisfy this equation yield a line with slope m that crosses the y-axis at $(0, b)$. In this equation, m and b may be any real numbers. This form does not account for vertical lines. However, the equation $x = a$ yields a vertical line that crosses the x-axis at $(a, 0)$.

A more general form of the equation of a line in the plane is

$$ax + by + c = 0. \tag{6.1}$$

This equation has two minor caveats.

> **Lines from the equation $ax + by + c = 0$ in other forms**
>
> Equation (6.1) can be converted to other forms. For example, we can solve for y like this:
>
> $$ax + by + c = 0$$
> $$by = -ax - c$$
> $$y = \frac{-a}{b}x - \frac{c}{b}$$
>
> where the last step is valid provided $b \neq 0$. In this case, Equation (6.1) represents a line with slope $-a/b$.
>
> In case $b = 0$, it must be the case that $a \neq 0$. Then we can solve for x:
>
> $$ax + c = 0$$
> $$ax = -c$$
> $$x = -c/a.$$
>
> In this case, Equation (6.1) represents a vertical line.

- First, we do not allow $a = b = 0$ because either there would be no solutions (if $c \neq 0$) or every point in the plane would satisfy (6.1) (if $c = 0$).

- Second, the same line may be represented by more than one equation. For example, the line $2x - 3y + 5 = 0$ is the same as $4x - 6y + 10 = 0$ because the second equation is simply a multiple of the first.

We use the notation $[a, b, c]$ to stand for the line determined by (6.1). The caveats tell us that we may not have $a = b = 0$ and that the line $[a, b, c]$ is the same as the line $[ma, mb, mc]$ provided $m \neq 0$.

The notation $[a, b, c]$ is a *homogeneous* representation of the line. The term *homogeneous* in this context means that all nonzero multiples of $[a, b, c]$ give the exact same line.

Given two lines $[a, b, c]$ and $[d, e, f]$, we use algebra to determine their point of intersection. For example, consider the lines $[1, 2, 3]$ and

[4, 5, 6]. To find their intersection, we solve this pair of equations:

$$1x + 2y + 3 = 0$$
$$4x + 5y + 6 = 0$$

If we multiply the first of these by 4 and subtract the second, we can solve for y:

$$4x + 8y + 12 = 0$$
$$- \quad 4x + 5y + 6 = 0$$
$$\overline{3y + 6 = 0}$$
$$\therefore \quad y = -2$$

Substituting $y = -2$ into $1x + 2y + 3 = 0$ gives $x - 4 + 3 = 0$ and so $x = 1$. Therefore, lines $[1, 2, 3]$ and $[4, 5, 6]$ intersect at $(1, -2)$.

Now consider the lines $[1, 2, 3]$ and $[1, 2, 4]$. These are different lines, but unsurprisingly, there is no solution to the equations

$$1x + 2y + 3 = 0$$
$$1x + 2y + 4 = 0$$

because these lines are parallel. However, when we extend our frame of reference to the projective plane, we hope to find a point "at infinity" that is on both of these lines. To do this, we create a new way to assign coordinates to points.

Homogeneous coordinates for points

The standard way to represent a point in the plane is (x, y) where x and y are real numbers. However, this notation is not useful for describing points "at infinity." The innovation is to use *homogeneous coordinates* for points. In this new notation, the point (x, y) becomes $(x, y, 1)$. More than that, any nonzero multiple of this triple names the same point.

For example, the point $(3, -2)$ is represented in this new notation as $(3, -2, 1)$, and all of the following are ways of naming the exact same point:

$$(30, -20, 10), \quad (-6, 4, -2), \quad \text{and} \quad (\tfrac{3}{2}, -1, \tfrac{1}{2}).$$

Because nonzero multiples of (x, y, z) all stand for the same point, we call this notation *homogeneous coordinates*.

With this modification, line $[a, b, c]$ is now represented by this equation:

$$ax + by + cz = 0. \tag{6.2}$$

This is a minor modification of (6.1) as we transit from standard (x, y) coordinates to homogeneous coordinates.

Notice that if (x, y) satisfies $ax + by + c = 0$, then (of course) $(x, y, 1)$ satisfies $ax + by + cz = 0$. Further, any nonzero multiple of $(x, y, 1)$—say, (mx, my, mz)—also satisfies (6.2):

$$a(mx) + b(my) + c(mz) = m(ax + by + cz) = 0.$$

Let's consider parallel lines $[1, 2, 3]$ and $[1, 2, 4]$. Where do they meet? The new equations to solve are these:

$$1x + 2y + 3z = 0$$
$$1x + 2y + 4z = 0$$

We can solve these equations. First of all, subtracting one equation from the other gives $z = 0$. This reduces the pair of equations to the single $x + 2y = 0$. This equation has many solutions such as $(-2, 1)$ or $(2, -1)$ or $(10, -5)$ and so on. Combining these with $z = 0$ gives the solution $(2, -1, 0)$ or any nonzero multiple of $(2, -1, 0)$.

Notice that the last coordinate of $(2, -1, 0)$ is zero. This can't be an "ordinary" point in the plane, because all those points are nonzero multiples of $(x, y, 1)$. Because $(2, -1, 0)$ is the point of intersection of parallel lines, it is a point at infinity!

We now expand our understanding of homogeneous coordinates. Points in the projective plane are represented by a triple of the form (x, y, z) in which we disallow $x = y = z = 0$. Any nonzero multiple of (x, y, z) names the same point.

Ordinary points in the projective plane have $z \neq 0$, whereas points at infinity have $z = 0$.

We also gently expand the homogeneous representation of lines, $[a, b, c]$. Previously, we disallowed $a = b = 0$. Let's relax that and only forbid $a = b = c = 0$.

What does $[0, 0, 1]$ represent? It is the line that contains all points of the form (x, y, z) such that

$$0x + 0y + 1z = 0 \quad \Rightarrow \quad z = 0$$

—that is, all points of the form $(x, y, 0)$. Line $[0, 0, 1]$ is the line at infinity!

Duality revisited

By representing points in the projective plane as homogeneous coordinates (x, y, z) and lines as $[a, b, c]$, we have no special cases for any points or lines.

The criteria that point (x, y, z) sits on line $[a, b, c]$ is given the simple equation $ax + by + cz = 0$. With this notation, the only difference between points and lines is the shape of the brackets in their notation. Algebraically, and hence geometrically, points and lines behave exactly like each other.

For example, earlier we had this pair of statements:

- *Given any two distinct points, there is a unique line that contains them both.*

- *Given any two lines, there is a unique point contained in both of them.*

The first says, if (x_1, y_1, z_1) and (x_2, y_2, z_2) are distinct points, then there is a line $[a, b, c]$ such that $ax_i + by_i + cz_i = 0$ for both $i = 1$ and $i = 2$.

The second says, if $[a_1, b_1, c_1]$ and $[a_2, b_2, c_2]$ are distinct lines, then there is a point (x, y, z) such that $a_i x + b_i y + c_i z = 0$ for $i = 1$ and $i = 2$.

These are exactly the same, once we swap letters x, y, z for letters a, b, c.

6.4 Desargues's theorem

A beautiful theorem of projective geometry is found in the work of the seventeenth-century French mathematician Girard Desargues.

Suppose we have two triangles ABC and DEF in which the corresponding sides are parallel. That is, $AB \| DE$, $AC \| DF$, and $BC \| EF$ (see Figure 6.4). When we connect corresponding points in the two triangles, the three lines AD, BE, and CF are concurrent;[2] they all contain the point Q.

Swap the roles of points and lines. Instead of requiring the *sides* of the two triangles to determine parallel lines, we suppose that *vertices* determine parallel lines. That is, we suppose that the three lines AD, BE,

---

[2] We call a collection of lines *concurrent* if they all contain a common point. It is the dual notion to *collinear*.

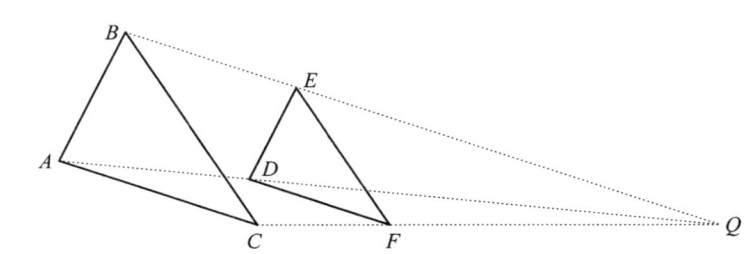

Figure 6.4: Triangles ABC and DEF have the property that $AB \parallel DE$, $AC \parallel DF$, and $BC \parallel EF$. The lines AD, BE, and CF meet at a common point Q.

and CF are parallel. Extending the edges in pairs determines three points. That is, AB and DE intersect at X, BC and EF intersect at Y, and AC and DF intersect at Z. The conclusion is that the three points of intersection, X, Y, and Z, are collinear, as illustrated in Figure 6.5.

We have been intentionally vague about whether we are in the Euclidean or projective plane. In fact, it doesn't matter because when we speak of lines being parallel, we simply mean that their common point of intersection is at infinity.

You may have noticed that in Figure 6.4 the two triangles are different sizes (but because their sides are parallel, they are similar triangles). What happens if ABC and DEF are actually congruent triangles with corresponding sides parallel? In that case, the lines AD, BE, and CF are parallel and their common point of intersection, Q, lies at infinity.

Likewise, in Figure 6.5 it is possible that some corresponding sides of the triangles are parallel. This is not a problem! In this case some (or all) of the points X, Y, and Z are at infinity, and are still collinear.

In the projective plane, there is nothing special about points at infinity or the line at infinity. Points at infinity simply have coordinates of the form $(x, y, 0)$, and the line at infinity has coordinates $[0, 0, 1]$. With this in mind, let's revisit the situation illustrated in Figure 6.4.

Instead of saying that lines AB and DE are parallel, we just say that they meet at X. Perhaps X is at infinity, or perhaps not. Likewise, suppose that BC and EF meet at Y, and that AC and DF meet at Z.

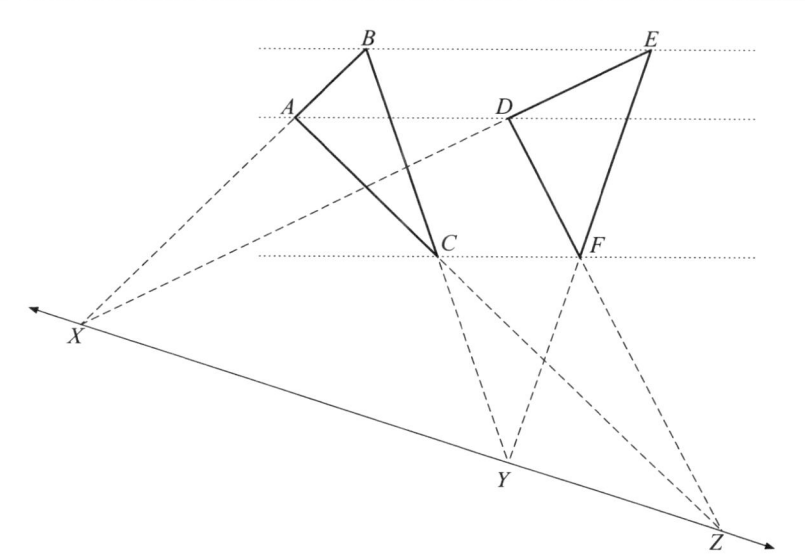

Figure 6.5: Triangles ABC and DEF have the property that lines AD, BE, and CF are parallel. Let X be the intersection of lines AB and DE, Y be the intersection of BC and EF, and Z be the intersection of AC and DF. The result is that the three points X, Y, and Z are collinear. This result is dual to the one illustrated in Figure 6.4, in which collinearity has been replaced by concurrency.

One assertion of Desargues's theorem is that if X, Y, and Z are collinear, then the lines AD, BE, and CF are concurrent: they meet at a point Q. Furthermore, Desargues's theorem asserts the converse. If lines AD, BE, and CF are concurrent, then the points X, Y, and Z must be collinear. All this is illustrated in Figure 6.6.

Here is a technical way to express Desargues's theorem: *Two triangles are in perspective axially if and only if they are in perspective centrally.* The phrase *in perspective axially* refers to the collinearity of the three points X, Y, and Z. Their common line is called the *axis of perspective*. The phrase *in perspective centrally* refers to the concurrency of the lines AD, BE, and CF. Their common point of intersection, Q, is called the *center of perspectivity*.

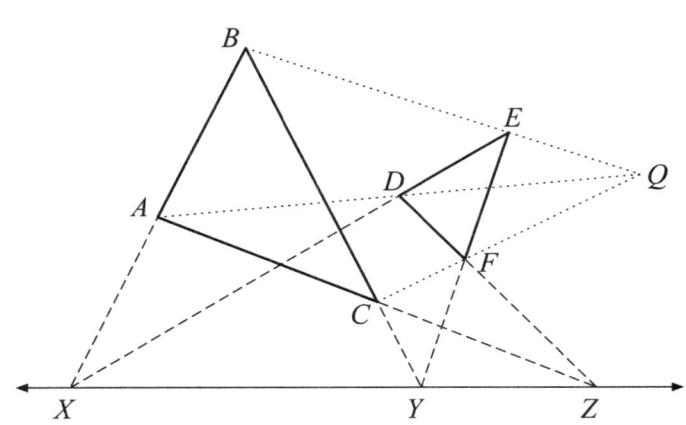

Figure 6.6: In this figure, lines AB and DE intersect at X, BC and EF intersect at Y, and AC and DF intersect at Z. Desargues's theorem asserts that points X, Y, and Z are collinear if and only if lines AD, BE, and CF are concurrent.

6.5 Topology of the projective plane

Topology is the branch of mathematics that studies properties of geometric shapes that are unaffected by smooth transformations. In topology, a circle and an ellipse are considered to be equivalent because one can be deformed into the other by stretching. People traversing a circular path net one full rotation over the course of their outing. As they walk, they are continuously turning until they net a 360° turn. The same is true for an elliptical path.

However, if the path has the shape of the infinity symbol, ∞, then walkers will begin turning in one direction and then eventually reverse their rotation. In the end, they net a 0° rotation.

We know that the projective and ordinary (Euclidean) planes are geometrically different—in the first all pairs of lines intersect, but in the second some pairs of lines do not. In this section we show that they are topologically different.

Homogeneous coordinates in a standard form

Points in the projective plane are named using homogeneous coordinates, (x, y, z). The only triple that isn't the name of a point is $(0, 0, 0)$. A feature of homogeneous coordinates is that any nonzero multiple of a triple names the same point. So the point $(2, -1, 2)$ is the same as the point $(-1, \frac{1}{2}, -1)$ and $(20, -10, 20)$ and so forth.

Let's consider a way to choose a "standard" name for a point in the projective plane. There are various choices, but the one we consider here is to choose the coordinates (x, y, z) so that

$$x^2 + y^2 + z^2 = 1. \tag{6.3}$$

This standardization greatly reduces the options for a point. For the point $(2, -1, 2)$ we note that $2^2 + (-1)^2 + 2^2 = 9$, so if we multiply all the coordinates by $\frac{1}{3}$ (because $3 = \sqrt{9}$), we have the equivalent triple $(\frac{2}{3}, -\frac{1}{3}, \frac{2}{3})$, which satisfies the desired requirement: $(\frac{2}{3})^2 + (-\frac{1}{3})^2 + (\frac{2}{3})^2 = 1$.

Equation (6.3) doesn't completely standardize the choice for the coordinates because $(2, -1, 2)$ is also equivalent to $(-\frac{2}{3}, \frac{1}{3}, -\frac{2}{3})$ and it, too, satisfies (6.3).

Given the choice between coordinates (x, y, z) and $(-x, -y, -z)$ satisfying (6.3), let's choose the one in which $z > 0$ is the standard choice. Thus, the standard choice for point $(2, -1, 2)$ is $(\frac{2}{3}, -\frac{1}{3}, \frac{2}{3})$.

Alas, this approach doesn't work for those points at infinity because their third coordinate is 0. However, for the discussion we are about to present, we are content with points at infinity having two standard forms.

Note that if a point's coordinates (x, y, z) are standardized, we can determine z just from knowing x and y because (6.3) can be rearranged like this:

$$z = \sqrt{1 - x^2 - y^2}.$$

Visualizing the projective plane

When put in standard form, every point in the projective plane satisfies equation (6.3). For ordinary points (x, y, z), which have $z \neq 0$, we see that

$$x^2 + y^2 + z^2 = 1 \quad \Rightarrow \quad x^2 + y^2 = 1 - z^2 < 1.$$

Points at infinity have $z = 0$, and in standard form we have $x^2 + y^2 = 1$.

We can see this geometrically. Given (x, y, z) in standard form, the point (x, y) is either in the interior of a circle of radius 1 centered at the

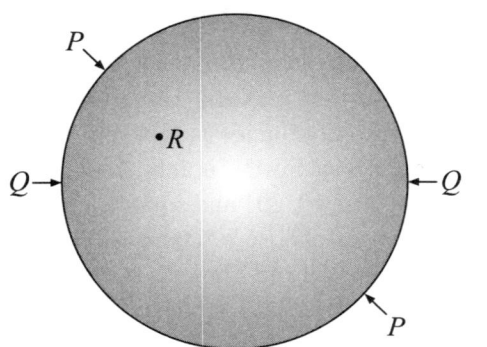

Figure 6.7: When a point (x, y, z) of the projective plane is in standard form, the coordinates (x, y) lie in the interior of a unit circle (if the point is an ordinary point) or on the boundary (if the point is at infinity). Ordinary points, such as R, have a unique representation, but points at infinity, such as P and Q, are present twice on the boundary.

origin (if (x, y, z) is an ordinary point), or else (x, y) is on the boundary of the unit circle (if the point is at infinity).

Ordinary points are uniquely determined by this method, but some ambiguity remains for points at infinity. For example, the point at infinity $(3, -4, 0)$ is represented by both of these diametrically opposite points on the unit circle: $(\frac{3}{5}, -\frac{4}{5})$ and $(-\frac{3}{5}, \frac{4}{5})$.

This is illustrated in Figure 6.7. Ordinary points appear in the interior of the disk. However, points at infinity appear twice on the boundary. This diagram should be interpreted such that diametrically opposite points on the boundary are the same.

Imagine a creature crawling about on the projective plane. If it heads toward the point Q on the far right and keeps going, it immediately emerges from the Q on the left. The two points marked Q in the diagram are the same.

The next step is to make a cut in the projective plane, as illustrated in Figure 6.8.

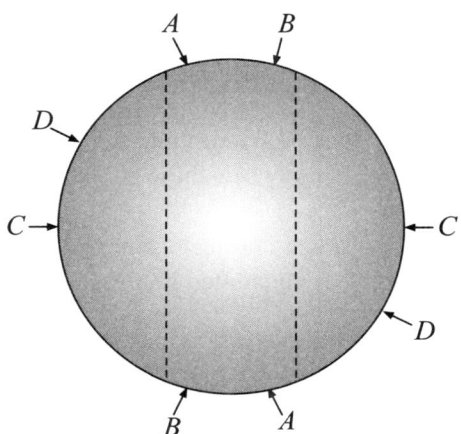

Figure 6.8: Cutting the projective plane along the dashed path decomposes the projective plane into a Möbius band and a disk.

We cut along the dashed path. At first glance, it appears that we are making two cuts. However, a closer examination reveals that this is a single cut because each point on the boundary is the same as the point located diametrically opposite. We start cutting at the top, just to the left of point A. We proceed downward until we reach the boundary point just to the left of the lower B. At this point we immediately continue from the top right (remember, the two Bs are the same!) and travel downward again. When we reach the point just to the right of the lower A, we have completed the cut.

Once the cut is complete, we are left with two pieces: the strip in the middle (between the dashed lines) and the rest. Again, it may appear that there are three pieces, but the lune-shaped region on the left is just part of the lune-shaped region on the right; points C and D on the boundary help us to visualize this. The strip we cut from the middle is a Möbius band because the top of the strip is flipped at the bottom.

The rest of the plane looks like a half loaf of pita bread. Imagine actually cutting out the two lune-shaped regions, but then realizing that the curved edges are common to both. That is, we sew the curved parts of the lunes to each other in a way that matches up the two points marked C, as well as matching up the two points marked D. If we now open up the pita bread (assuming it doesn't rip!) and stretch it out flat, we can make it into a disk in the plane.

Reversing this process, a projective plane can be assembled from a disk and a Möbius band. The disk has a single closed curve as its boundary, as does the Möbius band. Simply sew these two curves together, and you have a projective plane.

Unfortunately, such a sewing project cannot actually be completed in three-dimensional space, but undaunted knitters have tried.[3]

In closing, the fact that the projective plane contains a Möbius band but the Euclidean plane does not demonstrates that they are topologically different from each other.

6.6 Bézout's theorem

In Chapter 3 we introduced Bézout's theorem. If $p(x, y)$ is a polynomial of degree m and $q(x, y)$ is a polynomial of degree n, then the two curves $p(x, y) = 0$ and $q(x, y) = 0$ intersect in (at most) mn points.

This statement isn't exactly correct.

First, the algebraic curves $p(x, y) = 0$ and $q(x, y) = 0$ might overlap in infinitely many points—for example, if $p(x, y) = x^2 - y^2$ and $q(x, y) = x - y$. The two curves $p(x, y) = 0$ and $q(x, y) = 0$ include all points (x, y) with $x = y$. That's a minor issue that is easily remedied by this restatement.

> **Bézout's Theorem**
> If $p(x, y)$ is a polynomial of degree m and $q(x, y)$ is a polynomial of degree n, then if the curves $p(x, y) = 0$ and $q(x, y) = 0$ have finitely many intersections, then they intersect in mn points.

The careful reader will note that we dropped the parenthetical "at most." Hence, as presented, the statement is false. We need to do three things to make it true.

- First, count some intersections more than once.

---

[3] Readers can see some examples of knitted projective planes created by sarah-marie belcastro on her website www.toroidalsnark.net/mkpp.html.

- Second, use complex numbers.
- Third, work in projective geometry.

The first two ideas are familiar from working with single-variable polynomials. Let $p(x)$ be a polynomial of degree d. How many solutions does the equation $p(x) = 0$ have? The answer is d if we make some allowances.

First, suppose $p(x) = x^2 - 2x + 1$. Does the equation $p(x) = 0$ have two solutions? No: The only number x for which $p(x) = 0$ is $x = 1$. However, the answer is also—sort of—yes: The polynomial $p(x)$ factors $(x-1)(x-1)$. In this sense, $x = 1$ is a "double" root of the equation $p(x) = 0$. This is a reasonable allowance.

Second, suppose $p(x) = x^2 + 1$. How many solutions are there to the equation $p(x) = 0$? If we only allow real numbers, the answer is: There are none. However, it is reasonable to allow complex numbers, in which case the equation $p(x) = 0$ has two solutions: $x = i$ and $x = -i$.

With these two allowances (multiple counting and use of complex numbers), then, it is true that if $p(x)$ is a polynomial of degree d, then there are d solutions to the equation $p(x) = 0$.

This is thanks to the Fundamental Theorem of Algebra, which states that if $p(x)$ is a polynomial of degree d (whose coefficients are complex numbers), then $p(x)$ has a unique factorization of the form

$$p(x) = a(x - r_1)(x - r_2)\cdots(x - r_d)$$

where $a \neq 0$ and $r_1, r_2, \ldots, r_d$ are complex numbers. The equation $p(x) = 0$ has d solutions: $r_1, r_2, \ldots, r_d$.

Let's return to algebraic curves. Let

$$p(x, y) = x^2 + y^2 - 1 \quad \text{and} \quad q(x, y) = x^2 + 4y^2 - 1.$$

These curves can be seen in Figure 6.9.

Superficially, they appear to have only two points of intersection, but let's see how we might consider each of the intersections to be a double. Solve the equations $p(x, y) = 0$ and $q(x, y) = 0$ by subtracting one from the other to yield $3y^2 = 0$. This equation has 0 as a double root. Given $y = 0$, either equation yields $x = \pm 1$. Each y gives two xs.

Hence our four points of intersection are $(1, 0)$, $(1, 0)$, $(-1, 0)$, and $(-1, 0)$.

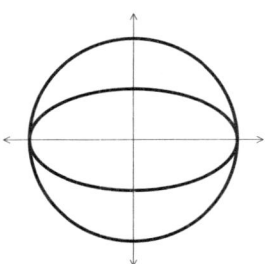

Figure 6.9: The algebraic curves $x^2 + y^2 - 1 = 0$ (the circle) and $x^2 + 4y^2 - 1 = 0$ (the ellipse) have two double intersection points.

Next consider these polynomials:

$$p(x,y) = x^2 + y^2 - 1 \quad \text{and} \quad q(x,y) = x + y - 2.$$

These are polynomials of degree 2 and 1; by Bézout's theorem, the curves $p(x,y) = 0$ and $q(x,y) = 0$ have $2 \cdot 1 = 2$ points of intersection.

Note that the curve $p(x,y) = 0$ is a radius-one circle centered at the origin and the curve $q(x,y) = 0$ is a line that lies entirely outside the circle! (This is illustrated in Figure 6.10.) They don't intersect at all. However, if we allow complex numbers, then we can solve the equations $p(x,y) = 0$ and $q(x,y) = 0$ to find

$$(x,y) = \left(1 - \frac{i\sqrt{2}}{2}, \frac{i\sqrt{2}}{2}\right) \quad \text{and} \quad (x,y) = \left(1 + \frac{i\sqrt{2}}{2}, -\frac{i\sqrt{2}}{2}\right)$$

are the two points of intersection.[4]

Now things get really interesting. The two polynomials we consider next are

$$p(x,y) = x^2 + y^2 - 1$$
$$\text{and} \quad q(x,y) = (x-2)^2 + (y-1)^2 - 1$$
$$= x^2 + y^2 - 4x - 2y + 4.$$

Since these polynomials both have degree two, we expect there to be four points of intersection.

---

[4]The algebra is not difficult.

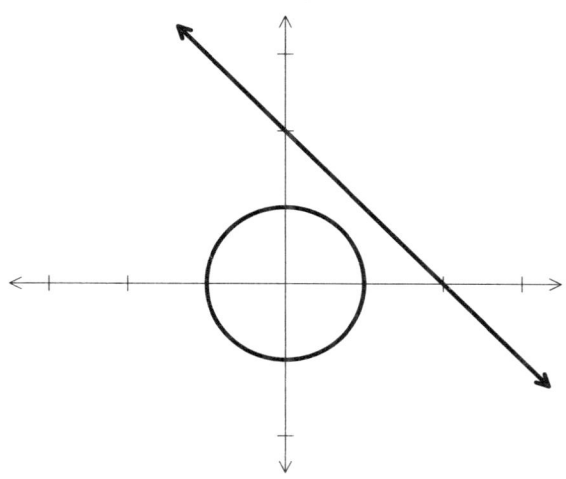

Figure 6.10: Curves $x^2 + y^2 - 1 = 0$ (the circle) and $x + y - 2 = 0$ (the line) do not appear to intersect at all.

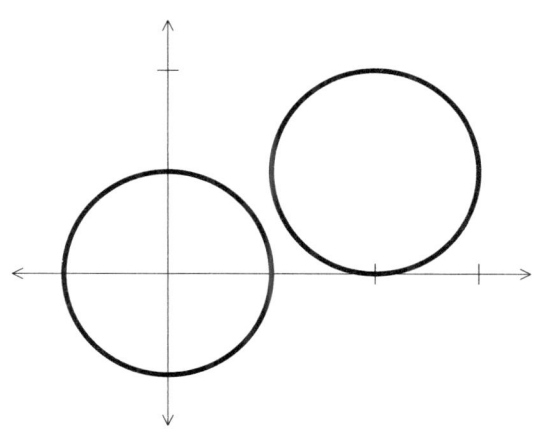

Figure 6.11: The curves $x^2 + y^2 - 1 = 0$ and $(x-2)^2 + (y-1)^2 - 1 = 0$ do not intersect. Or do they?

These are the equations of two radius-one circles: one centered at $(0,0)$ and the other at $(2,1)$. Geometrically, they do not intersect (Figure 6.11).

Subtracting the equation $q(x, y) = 0$ from $p(x, y) = 0$ eliminates the quadratic terms:

$$x^2 + y^2 - 1 = 0$$
$$- \quad x^2 + y^2 - 4x - 2y + 4 = 0$$
$$\overline{ 4x + 2y - 5 = 0.}$$

From this we have $y = (5-4x)/2$, which we substitute into $x^2 + y^2 - 1 = 0$:

$$x^2 + \left(\frac{5-4x}{2}\right)^2 - 1 = 0$$
$$\Rightarrow \quad 20x^2 - 40x + 21 = 0.$$

By the quadratic formula,

$$x = \frac{40 \pm \sqrt{40^2 - 4 \cdot 20 \cdot 21}}{2 \cdot 20} = 1 \pm \frac{i\sqrt{5}}{10}.$$

Using $y = (5 - 4x)/2$, we arrive at the following pair of points (in the complex plane) where these circles intersect:

$$\left(1 + \frac{i\sqrt{5}}{10}, \frac{1}{2} - \frac{i\sqrt{5}}{5}\right) \quad \text{and} \quad \left(1 - \frac{i\sqrt{5}}{10}, \frac{1}{2} + \frac{i\sqrt{5}}{5}\right).$$

We have found two points of intersection, but Bézout's theorem asserts that there are four. Where are they? Why, at infinity, of course!

The idea is to work in the *complex* projective plane, in which points and lines are homogeneous triples of complex numbers.

To find the other two points, we modify the equations $p(x, y) = 0$ and $q(x, y) = 0$ to work with homogeneous triples (x, y, z). The new versions of p and q are these:

$$p(x, y, z) = x^2 + y^2 - z^2$$
$$\text{and} \quad q(x, y, z) = (x - 2z)^2 + (y - z)^2 - z^2$$
$$= x^2 + y^2 - 4xy - 2yz + 4z^2.$$

Two important things to notice.

- First, when we evaluate $p(x, y, 1)$ and $q(x, y, 1)$, we get exactly the original polynomials.

- Second, all of the terms in the new polynomials have degree 2. This implies that if (x, y, z) satisfies $p(x, y, z) = 0$ [or $q(x, y, z) = 0$], then so does any nonzero multiple of (x, y, z). In other words, the equations $p(x, y, z) = 0$ and $q(x, y, z) = 0$ have been homogenized!

To find the missing two points of intersection, we look to points at infinity: points with homogeneous coordinates of the form $(x, y, 0)$. Plugging $(x, y, 0)$ into the equations gives

$$p(x, y, 0) = x^2 + y^2 + 0^2 = x^2 + y^2 = 0$$
$$\text{and} \quad q(x, y, 0) = (x-0)^2 + (y-0)^2 - 0^2 = x^2 + y^2 = 0.$$

The two equations are, quite simply, $x^2 + y^2 = 0$. We might be tempted to say, "Aha! We have $x = y = 0$." However, $(0, 0, 0)$ is forbidden. Nevertheless, there are other solutions when we permit complex numbers. From $x^2 = -y^2$ we find $x = \pm iy$. This leads to $(i, 1, 0)$ and $(-i, 1, 0)$ as points of intersection of $p(x, y, z) = 0$ and $q(x, y, z) = 0$. These last two points are at infinity—not in the *real* projective plane, but rather in the *complex* projective plane.

The complex projective plane is exactly what you might expect. The points are homogeneous triples (x, y, z) in which x, y, and z are complex numbers (not all zero). Likewise, lines are homogeneous triples $[a, b, c]$ as well. The point (x, y, z) lies on the line $[a, b, c]$ exactly when $ax + by + cz = 0$.

Summarizing: The original equations $p(x, y) = 0$ and $q(x, y) = 0$ have no intersection in the Euclidean plane. However, they have, indeed, four points of intersection in the complex projective plane:

$$\left(1 + \tfrac{i\sqrt{5}}{10}, \tfrac{1}{2} - \tfrac{i\sqrt{5}}{5}, 1\right), \left(1 - \tfrac{i\sqrt{5}}{10}, \tfrac{1}{2} + \tfrac{i\sqrt{5}}{5}, 1\right), (1, i, 0), \text{ and } (1, -i, 0).$$

We have "repaired" Bézout's theorem. Algebraic curves $p(x, y) = 0$ and $q(x, y) = 0$, given by polynomials of degree m and n, have exactly mn points of intersection, provided there are only finitely many intersections and that we appropriately count intersection multiplicity, use complex numbers, and allow points at infinity in the complex projective plane.

For your consideration

A point in the projective plane is a triple (x, y, z) (not all zero). Any nonzero multiple of (x, y, z) names the same point. In Section 6.5 we proposed a standardization for such triples by requiring $x^2 + y^2 + z^2 = 1$. This doesn't assign a unique triple to points because, for example, $(\frac{2}{3}, -\frac{1}{3}, \frac{2}{3})$ and $(-\frac{2}{3}, \frac{1}{3}, -\frac{2}{3})$ name the same point. Which triple would be the "better" name for the point in the projective plane? There isn't a reason to prefer one to the other.

Another way to think of a triple (x, y, z) is as a point in three-dimensional space. The requirement that $x^2 + y^2 + z^2 = 1$ places the point on a sphere of radius 1 centered at the origin—the unit sphere.

Now, if (x, y, z) is a point on the unit sphere, the only nonzero multiple of (x, y, z) that is also on the unit sphere is $(-x, -y, -z)$; this is the point diametrically opposite (x, y, z). We say that such points are *antipodal*.

This means that if p is a point on the unit sphere, and q is its antipodal point, then the pair $\{p, q\}$ is a reasonable way to name a point in the projective plane.

In this way, we can imagine each point in the projective plane corresponds to a pair of antipodal points on the unit sphere.

Also for your consideration: A line L in the projective plane is a set of points. Each point of L corresponds to two points on the sphere. Describe the entirety of all the points on the sphere arising from points on L.

Here is another, rather different thought for your consideration: The projective plane in this chapter is infinite. Since the coordinates are real numbers, this projective plane is called the *real projective plane*. There are, however, other projective planes, and here we describe a finite one.

The *Fano plane* consists of just seven points and seven lines. As in the real projective plane, points are named as triples (x, y, z) and lines are named as triples $[a, b, c]$, but in this case the only numbers we allow are 0s and 1s. As before, the triples are not permitted to be all zeros; we disallow $(0, 0, 0)$ as a point and $[0, 0, 0]$ as a line.

With these restrictions in place, there are seven points: $(0, 0, 1)$, $(0, 1, 0)$, $(0, 1, 1)$, $(1, 0, 0)$, $(1, 0, 1)$, $(1, 1, 0)$, and $(1, 1, 1)$. Similarly, there are seven lines.

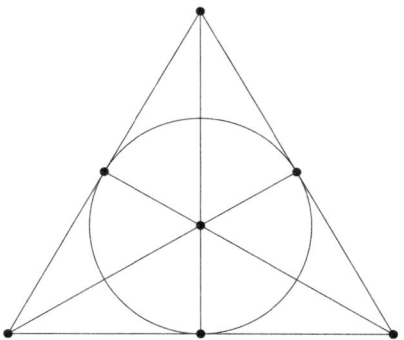

Figure 6.12: The Fano plane is a finite projective plane. The seven points are shown as dots. Six of the lines are shown as line segments, and one line is represented by the circle.

The rule for when a point (x, y, z) is on the line $[a, b, c]$ is that $ax + by + cz$ is even.[5] For example, the line $[1, 1, 0]$ contains these three points and no others: $(0, 0, 1)$, $(1, 1, 0)$, and $(1, 1, 1)$.

Some observations:

- Every line contains exactly three points.

- Every point is contained in exactly three lines.

- Given two different points, there is exactly one line that contains them both.

- Given two different lines, there is exactly one point common to them both.

Notice the duality between the first pair of statements and the second pair.

Finally, there is an attractive way to visualize the Fano plane. Figure 6.12 shows the seven points (represented by dots) and the seven lines (represented by six line segments and a circle joining three points each). Try to label the points and lines of this diagram using the triples (x, y, z) and $[a, b, c]$ (fourteen labels in all).

---

[5] A more sophisticated way to express this is that $ax + by + cz \equiv 0 \pmod{2}$.

Chapter 7

Line at Infinity II: Hyperbolic Plane

> An infinity of passion can be contained in one minute, like a crowd in a small space.
>
> —Gustave Flaubert, *Madame Bovary*

Let's face it: Infinity is big! So big that we cannot possibly wrap our arms around it. A plane has no edge; it goes on forever. We can never grasp it in its entirety.

However, in this chapter we present another geometry called the *hyperbolic plane*. We will see that this infinite world is entirely contained in a small space: a circle of radius one. Inside this circle we can create beautiful tilings that ignite artistic passion.

7.1 The parallel postulate

In (standard) Euclidean plane geometry, we derive various properties of figures, such as triangles or circles, from a handful of basic assumptions called *postulates* or *axioms*. For example, one postulate of plane geometry is that given two distinct points, there is a line that contains both of those points.

One might be tempted to think of axioms as statements that are obviously true and use them to prove more sophisticated theorems of plane geometry, such as the fact that the sum of the angles of a triangle is always $180°$.

84 ∞ Line at Infinity II: Hyperbolic Plane

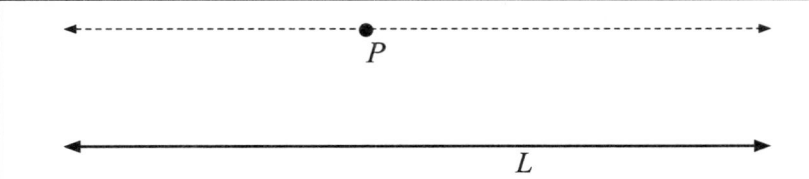

Figure 7.1: An illustration of the parallel postulate of Euclidean plane geometry:
Given a line L and a point P not on L, there is a unique line (shown dashed) containing P that does not intersect L.

This postulate does not hold for the projective plane (there can be no such line) nor in the hyperbolic plane (where there are at least two such lines).

An alternative way to think of the axioms of geometry is that they are definitions. Rather than start with definitions of points and lines, geometry is defined by its axioms; the axioms specify the properties the basic objects have, and that's good enough for us to work with them.

One of the axioms of plane geometry is the parallel postulate (illustrated in Figure 7.1):[1]

Parallel Postulate
Given a line L and a point P not on L, there is exactly one line that includes the point P that does not intersect L.

Note that *parallel lines* are defined as lines that do not intersect. Hence, the parallel postulate does not hold for the projective plane described in Chapter 6 because, in that setting, any two lines intersect (either at an ordinary point or at a point at infinity).

In other words, in the projective plane we can say this: Given a line L and a point P not on L, *every* line through P intersects L.

We now consider another alternative: We assume that given a line L and a point P not on L, there are *two* (or more) lines through P that do

---
[1] This postulate is actually due to John Playfair, an eighteenth-century Scottish mathematician. In the context of plane geometry, it is equivalent to Euclid's fifth postulate.

not intersect L. All the other requirements of geometry are unchanged. Is this possible?

Yes, indeed.

7.2 The hyperbolic plane

Points and lines

The basic objects of plane geometry are points and lines. There are foundational assumptions (known as axioms or postulates) about these concepts that determine the properties of plane geometry. One such basic property is that if P and Q are distinct points, then there is a unique line L that contains both P and Q.

The hyperbolic plane is a geometry that obeys the basic properties of ordinary (Euclidean) plane geometry, but in which the parallel postulate has been replaced by this axiom:

> Given a line L and a point P not on L, there are *two* (or more) lines containing P that do not intersect L.

A quick reaction might be: That's not possible! That understandable intuition comes from thinking about lines as being straight, but we can be more sophisticated in how we think about points and lines.

For Euclidean geometry, we can think of a point as a pair of real numbers (s, t). In this context, lines are equations of the form $ax+by+c = 0$ (where we disallow $a = b = 0$). To say that the point (s, t) is on the line $ax + by + c = 0$ simply means that $as + bt + c = 0$. There is no mention of the word "straight"; these points and lines satisfy the conditions of ordinary geometry. The fact that these notions of point and line obey the parallel postulate can be verified using algebra.

In Chapter 6 we had different concepts of points and lines. In the projective plane, a point is a triple of numbers (x, y, z) (not all zero) and a line is also a triple $[a, b, c]$ (not all zero). We also know that nonzero multiples of these triples stand for the same point or line. To determine if the point (x, y, z) is on line $[a, b, c]$, we simply check that $ax+by+cz = 0$.

The projective plane also has no notion of "straight." Points and lines in the projective plane obey basic properties of Euclidean geometry, but the parallel postulate does not hold.

> **Points and lines**
>
> What are points and lines in various types of plane geometry? This table gives a quick summary.
>
> | Type of plane | Point | Line |
> |---|---|---|
> | Euclidean | (s,t) | $ax + by + c = 0$ |
> | Projective | (x, y, z) homogeneous | $[a, b, c]$ homogeneous |
> | Hyperbolic | (s,t) with $s^2 + t^2 < 1$ | circular arc |

All this is to prepare for our introduction of the hyperbolic plane. In this setting, points and lines are, once again, different from our intuitive understanding.

The Poincaré disk model of the hyperbolic plane

To describe the hyperbolic plane, we need to specify what objects are its points and its lines. There are some choices in how to proceed, but a particularly attractive one is known as the *Poincaré disk*, named after the nineteenth-century French mathematician and physicist Henri Poincaré.

The *points* of the hyperbolic plane are pairs of real numbers, (s,t), with the additional requirement that $s^2 + t^2 < 1$. In other words, the points of the hyperbolic plane are "standard" points that sit in the interior of a circle of radius 1 centered at the origin.

The *lines* of the hyperbolic plane are circular arcs that are perpendicular to the unit circle as well as diameters of the unit circle.

The hyperbolic plane and some of its lines are shown in Figure 7.2. The entire hyperbolic plane fits inside the unit circle (shown dashed in the diagram). All points of the hyperbolic plane are interior to that circle; points on the circle are legitimately considered to be "at infinity" and are not part of the hyperbolic plane. They are referred to as *ideal points*.[2]

---

[2]Recall that the line at infinity in the projective plane (see Chapter 6) is part of the projective plane. In the hyperbolic plane, the ideal points are not part of the hyperbolic plane.

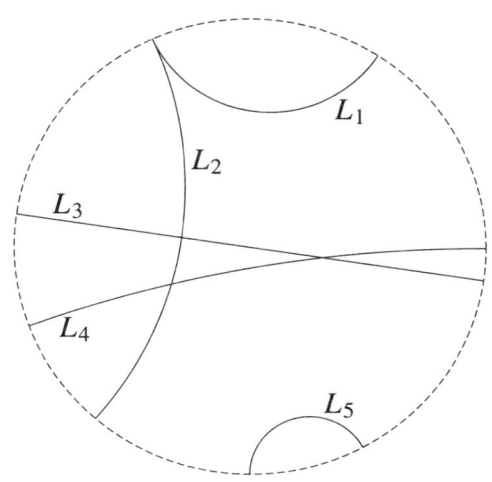

Figure 7.2: Some lines in the hyperbolic plane. Lines L_1 and L_2 do not intersect, but can be imagined to meet at infinity—the boundary of the hyperbolic plane.

We should note that lines L_1 and L_2 do not intersect, although they can be said to meet at infinity. By contrast, lines L_4 and L_5 do not intersect and do not meet at infinity.

Line L_3 passes through the center of the circle, and it is rendered as a line segment joining diametrically opposite points on the circle. The other lines are shown as circular arcs that meet the dashed circle at 90°.

How big is the hyperbolic plane?

Given that all points in the hyperbolic plane are in the interior of a circle, and looking at Figure 7.2, one might wonder: Is the hyperbolic plane finite? Since the center of the disk is distance 1 from the dashed circle, are all distances between points in the hyperbolic plane modest (less than 2, which would be the length of a diameter)?

In fact, the hyperbolic plane is infinite and the distance from any point in the interior to the dashed line is infinite, but instead of using the usual Euclidean distance, we use a different distance function. See Section 7.4.

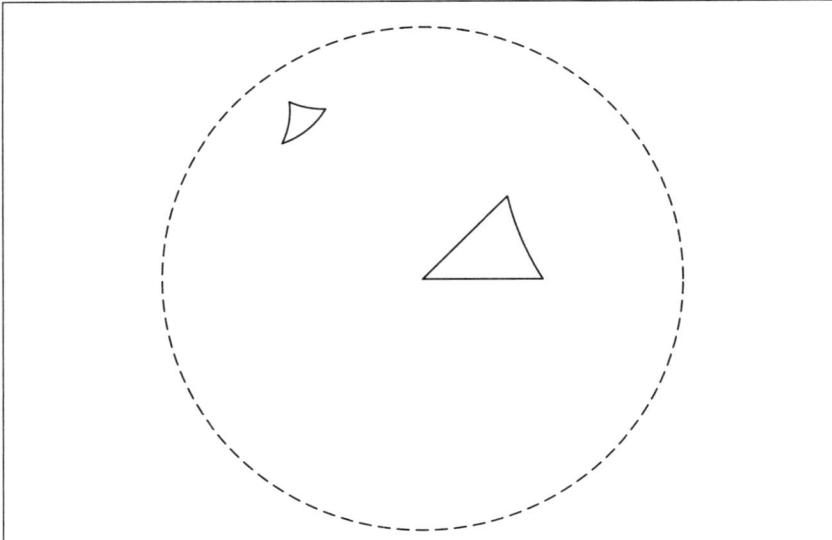

Figure 7.3: The two triangles shown in this picture of the hyperbolic plane are congruent; they are exactly the same size and shape. The Poincaré disk model can be thought of as a fish-eye lens that shows the entirety of the hyperbolic plane inside a circle. Items farther from the center are made smaller (in order to fit into the circle).

Deriving the distance function from the basic properties of the hyperbolic plane is rather complicated, but the upshot is that distances between points that appear close together dramatically increase as one gets closer to the line at infinity (the dashed circle).

For example, the two triangles in Figure 7.3 are exactly the same size and shape. The triangle closer to the line at infinity only looks smaller in the picture. Another example of how the disk model of the hyperbolic plane acts like a fish-eye lens is shown in Figure 7.4.

A revised parallel postulate

The parallel postulate does not hold for the hyperbolic plane. This is illustrated in Figure 7.5, in which we show a line L and a point P that is not on L. There are two lines that contain P that do not intersect L.

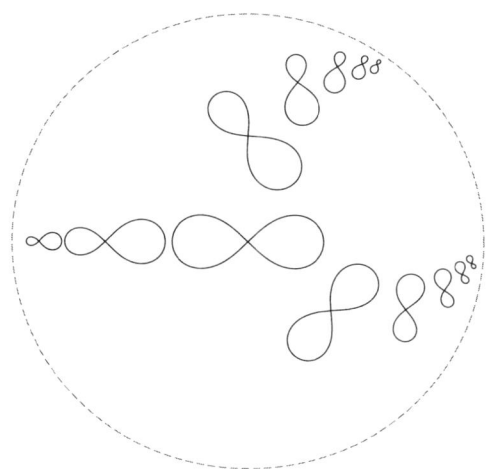

Figure 7.4: All the curves in this image of the hyperbolic plane are exactly the same size and shape.

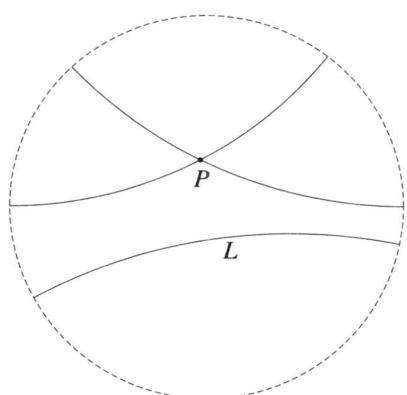

Figure 7.5: A line L and a point P that is not on L lying in the hyperbolic plane. Observe that there are two lines that contain P that do not intersect L.

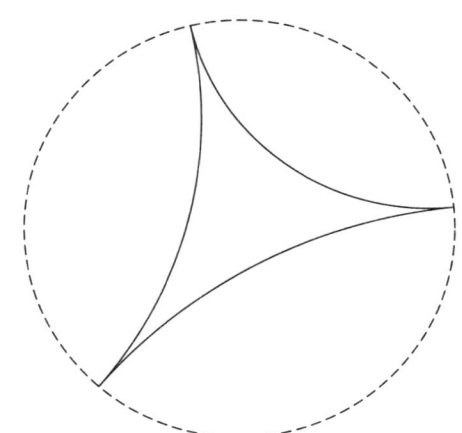

Figure 7.6: This triangle in the hyperbolic plane has its three vertices at infinity. The sum of the angles of any triangle in the hyperbolic plane is always less than 180° (π radians); the angle sum for this triangle is 0.

More dramatic still, one can see that there are *infinitely* many lines through P that do not intersect L.

There is an interesting consequence of this property concerning triangles. In ordinary, Euclidean geometry, one of the propositions one proves about triangles concerns the sum of their angles: the result is always 180° (π radians). The proof of this begins as follows: *Given triangle ABC, draw a line through C that is parallel to AB*. This is unambiguous in Euclidean geometry, but not in the hyperbolic plane. A consequence of hyperbolic geometry is that the sum of the angles of any triangle is always less than 180° (less than π). How much less? In the extreme, if we make a triangle whose vertices are at three different ideal points, then the sum of the angles is 0. This is illustrated in Figure 7.6.

Upper half-plane model

The Poincaré disk model of the hyperbolic plane is aesthetically pleasing because the entire plane can be visualized in the interior of a circle. There

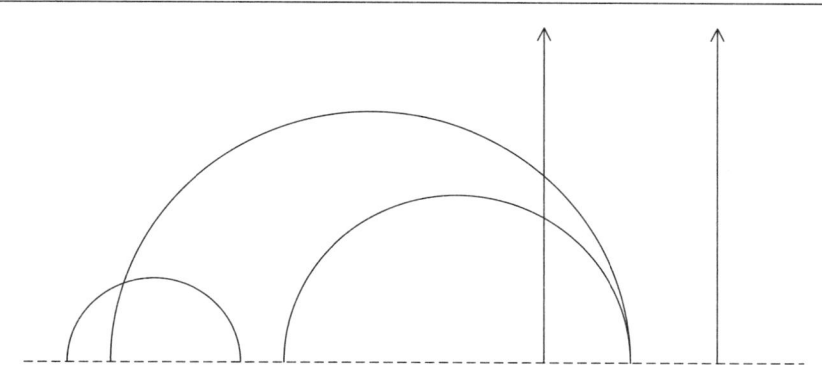

Figure 7.7: The upper half-plane model of the projective plane is an alternative to the Poincaré disk model. The points of the hyperbolic plane are the points (x, y) with $y > 0$. A line in this model is either a semicircle centered on the x-axis, or a vertical ray. The x-axis (shown dashed) is the line at infinity.

are other ways to visualize the hyperbolic plane, including the *upper half-plane* model.

In this model, the *points* of the hyperbolic plane are all the points in the usual Euclidean plane that lie strictly above the x-axis. The *lines* of the hyperbolic plane are either semicircles centered on the x-axis or vertical rays starting on the x-axis. This is illustrated in Figure 7.7.

In this setting, the x-axis becomes the line at infinity for the hyperbolic plane.

There is an algebraic way to relate the two models: upper half-plane and Poincaré disk. To make this connection, we use complex numbers. As in Chapter 5, points in the plane can be thought of as complex numbers: $(a, b) \leftrightarrow a + bi$; see Figure 5.1. Using this correspondence, a complex number $z = a + bi$ is a point in the Poincaré disk model of the hyperbolic plane, provided $a^2 + b^2 < 1$. Likewise, $z = a + bi$ is a point in the upper half-plane model, provided $b > 0$.

(Recall that the line at infinity—either the circle bounding disk or the x-axis, depending on the model—is *not* part of the hyperbolic plane.

Lines in the hyperbolic plane connect two different ideal points, but those ideal points are not part of the line.)

The translation between the two models uses the following function:[3]

$$f(z) = \frac{z-i}{z+i}.$$

Specifically, if z is a point in the upper half-plane, then $f(z)$ is a point in the interior of the unit circle. Further, each line in the upper half-plane model is converted by f to a line in the Poincaré disk model.

For example, some of the lines in the upper half-plane model are vertical rays emanating from the x-axis. Such lines join two ideal points: one on the x-axis and one at (complex) infinity. How does f transform those lines?

By convention (see Section 5.4), $f(\infty) = 1$. For points on the real axis, $z = a + 0i$, we have

$$f(z) = f(a + 0i) = \frac{a-i}{a+i} = \left[\frac{a-i}{a+i}\right] \cdot \left[\frac{a-i}{a-i}\right] = \frac{(a-i)^2}{a^2+1}$$

$$= \frac{a^2 - 1 - 2ai}{a^2 + 1} = \left(\frac{a^2-1}{a^2+1}\right) + \left(\frac{-2a}{a^2+1}\right)i.$$

We need to verify that $f(z)$ is an ideal point. That is, writing $f(z) = c + di$, we need to show that $c^2 + d^2 = 1$ (and so $f(z)$ lies on the unit circle). Here are the calculations:

$$\left(\frac{a^2-1}{a^2+1}\right)^2 + \left(\frac{-2a}{a^2+1}\right)^2 = \frac{(a^2-1)^2 + (-2a)^2}{(a^2+1)^2}$$

$$= \frac{(a^4 - 2a^2 + 1) + 4a^2}{a^4 + 2a^2 + 1} = \frac{a^4 + 2a^2 + 1}{a^4 + 2a^2 + 1} = 1.$$

In words, the vertical lines in the upper half-plane model are transformed by f to lines in the Poincaré disk model that join the ideal point $1 + 0i$ to other ideal points on the line at infinity (the unit circle), as shown in Figure 7.8.

7.3 Tessellations

A *tessellation* is a tiling of the plane by polygonal shapes. In this section, our focus is on tessellations whose tiles are regular polygons—polygons

---
[3] This function is a linear fractional transformation. These types of functions are described in Section 5.4.

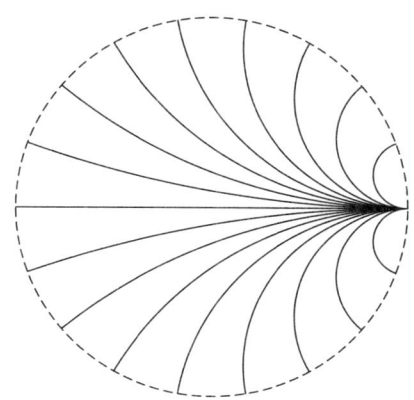

Figure 7.8: The vertical lines in the upper half-plane model of the hyperbolic plane are transformed by the function $f(z) = (z-i)/(z+i)$ to be the lines in the Poincaré disk model that connect the ideal point $1 + 0i$ to the other ideal points.

whose edges are all the same length and whose angles are all equal to each other. Further, we require that the corners of a tile line up exactly with the corners of adjacent tiles.

With these restrictions, there are only three ways to tile the Euclidean plane with regular polygons: equilateral triangles, squares, and regular hexagons. These familiar tessellations are illustrated in Figure 7.9.

It's not possible to tile the Euclidean plane with regular pentagons because each corner of a regular pentagon is 108 degrees, and 360 is not divisible by 108. A similar analysis shows that it is not possible to tile the plane with regular n-gons except for n = 3, 4, or 6.

The situation in the hyperbolic plane is much more interesting. As we noted, the angles of a triangle sum to less than 360°. The larger the triangle, the smaller its angles. By carefully choosing the side length of an equilateral triangle, we can set its angles to be any size less than 60°. For example, we can choose the angle size to be 360/7 degrees, which enables us to tile the hyperbolic plane with equilateral triangles in which seven triangles meet at each corner. Likewise, we can take somewhat larger triangles whose angles are all 360/8 degrees, and tile

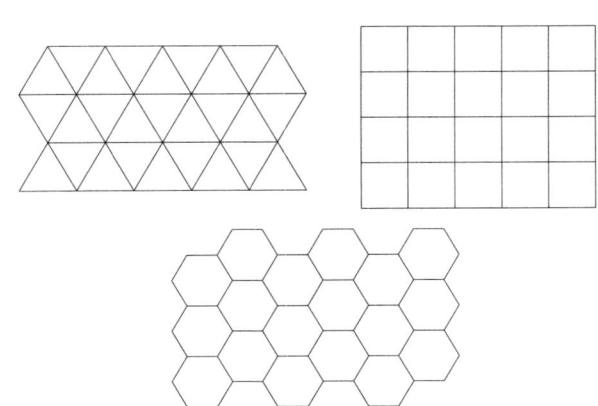

Figure 7.9: The only regular polygons that tile the Euclidean plane are triangles, squares, and hexagons.

the hyperbolic plane with equilateral triangles in which eight triangles meet at each corner. This is beautifully illustrated in Figure 7.10.

The triangles in the diagram appear to be larger in the center than in the periphery, but in fact all the triangles in one of these tessellations are the same size as each other.

In the Euclidean plane, a square's angles are all 90°. In the hyperbolic plane, a regular quadrilateral's angles are all smaller than 90°. That makes it impossible to tile the hyperbolic plane so that four regular quadrilaterals meet at each corner, but we can create tessellations in which five or more meet at the corners. This is illustrated in Figure 7.11.

The Euclidean plane cannot be tiled with regular pentagons, but the hyperbolic plane can! Some examples are shown in Figure 7.12.

Similarly, we cannot create a tessellation of the hyperbolic plane with regular hexagons in such a way that three hexagons meet at each corner, but we can create such tilings in which four (or more) meet at the corners (Figure 7.13).

We need not stop at hexagons. The hyperbolic plane can be tiled by regular n-gons for all values of $n \geq 3$. Examples with $n = 7$ and $n = 8$ are shown in Figure 7.14.

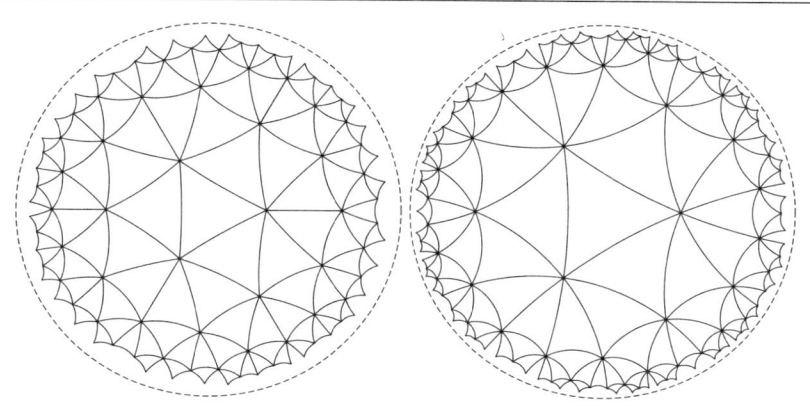

Figure 7.10: Tiling the hyperbolic plane with regular triangles. In the tiling on the left, seven triangles meet at each corner. On the right, eight triangles meet at each corner.

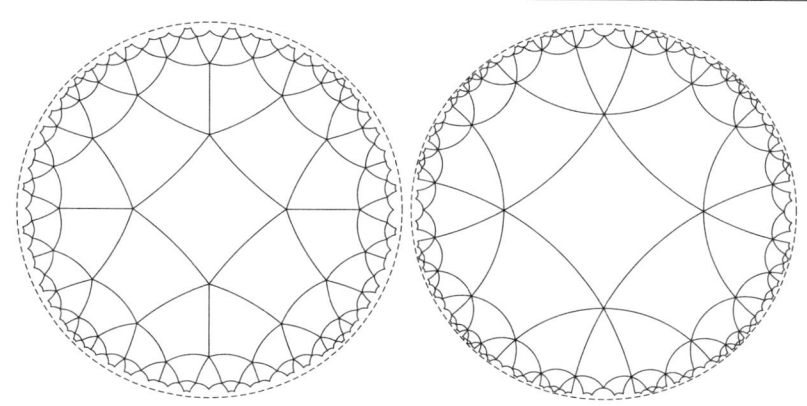

Figure 7.11: Tiling the hyperbolic plane with regular quadrilaterals. In the tiling on the left, five quadrilaterals meet at each corner. On the right, six quadrilaterals meet at each corner.

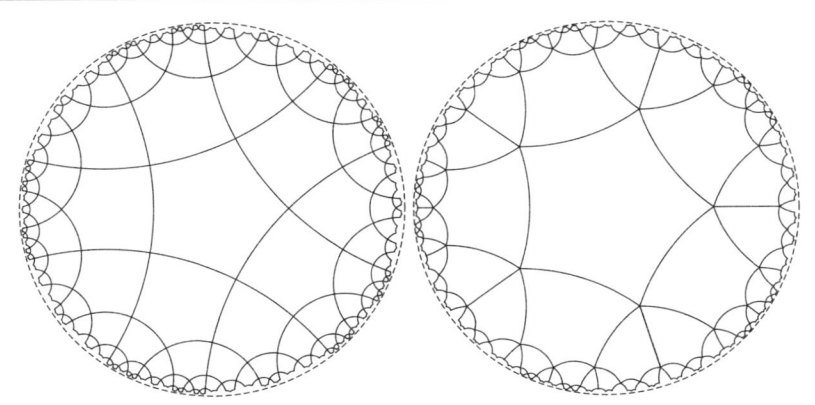

Figure 7.12: Tiling the hyperbolic plane with regular pentagons. In the tiling on the left, four pentagons meet at each corner. On the right, five pentagons meet at each corner.

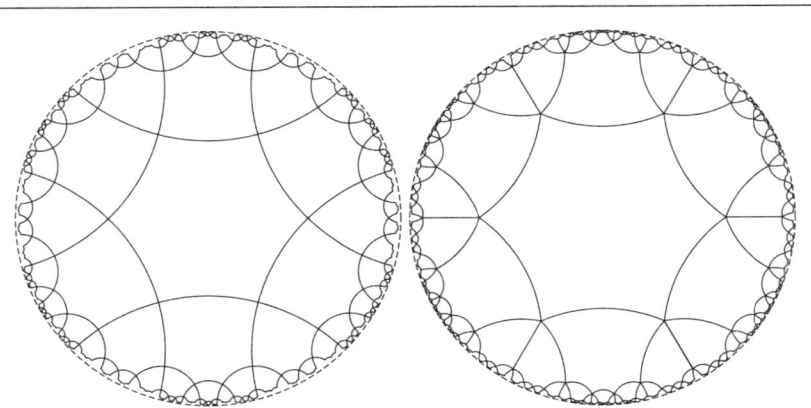

Figure 7.13: Tiling the hyperbolic plane with regular hexagons. In the tiling on the left, four hexagons meet at each corner. On the right, five hexagons meet at each corner.

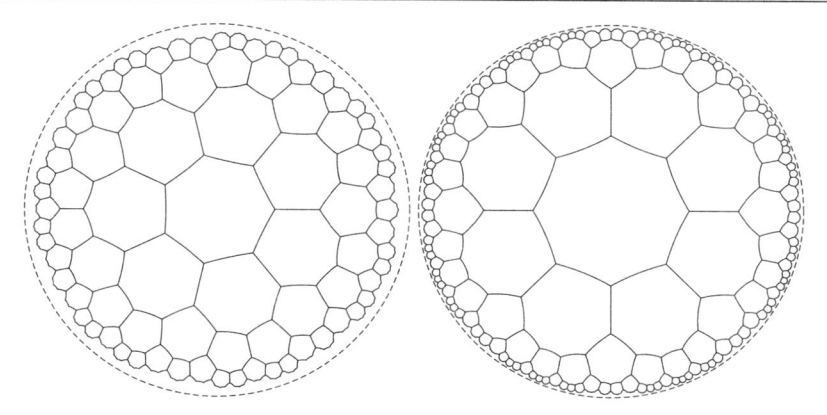

Figure 7.14: Tiling the hyperbolic plane with regular heptagons and octagons. In both cases, three polygons meet at each corner.

Tessellations of the hyperbolic plane are beautiful unto themselves, but become even more delightful when incorporated into works of art, as in the *Circle Limit* woodcuts by M. C. Escher. For example, *Circle Limit I* is based on a tiling of the hyperbolic plane by regular quadrilaterals, whereas *Circle Limit III* combines regular triangles and quadrilaterals.[4]

7.4 Distance and area in the hyperbolic plane

Distance between points

In the Euclidean plane, the distance between points (a, b) and (c, d) can be deduced from the Pythagorean theorem; the distance between them is

$$\sqrt{(a-c)^2 + (b-d)^2}.$$

The situation in the hyperbolic plane is more complicated. Although there are right triangles in the hyperbolic plane, they do not satisfy the Pythagorean theorem. To begin, consider the two points $z = a + bi$ and

---

[4]Images of these works by Escher can be found at mathstat.slu.edu/escher/index.php/Circle_Limit_Exploration.

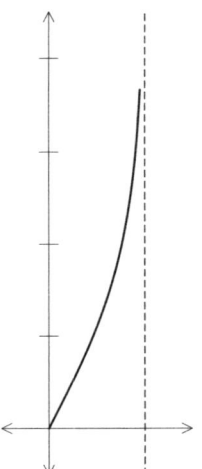

Figure 7.15: Graph showing the distance in the hyperbolic plane between the origin, $0 + 0i$, and a point $a + bi$ in the unit disk whose Euclidean distance from the origin is r. The distance between these points in the hyperbolic plane is given by the formula $\log[(1+r)/(1-r)]$.

In this graph, the horizontal axis is r and the vertical axis is the distance in the hyperbolic plane.

$0 = 0 + 0i$ (the center of the disk). The *Euclidean* distance between these points is

$$r = \sqrt{a^2 + b^2}.$$

This is not the distance in the hyperbolic plane! That distance is

$$\log\left(\frac{1+r}{1-r}\right) \tag{7.1}$$

(where the base of the logarithm is e). A graph of equation (7.1) is shown in Figure 7.15.

Because the point $a + bi$ is in the interior of the disk, its Euclidean distance, r, from the origin $0 + 0i$ is less than 1. As r approaches 1, the hyperbolic plane distance from the origin gets exceedingly large, approaching infinity.

A more complicated formula may be derived for two arbitrary points $a + bi$ and $c + di$ in the interior of the circle.

Area of a triangle

In Euclidean geometry, the area of a triangle is given by the familiar formula $\frac{1}{2}bh$ where b is the length of the side connecting two corners of the triangle (the "base") and h is the distance from the base to the third corner (the "height").

There are several other formulas for the area of a triangle such as Heron's formula: Let a, b, and c be the lengths of the sides of the triangle. Let $s = (a + b + c)/2$, that is, half the perimeter of the triangle. Then Heron's formula for the area of the triangle is

$$\sqrt{s(s-a)(s-b)(s-c)}.$$

One more: If a and b are the lengths of two sides of a triangle, and θ is the size of the angle formed by those sides, then the area of the triangle is given by this expression:

$$\tfrac{1}{2}ab \sin \theta.$$

Amazingly, the formula for the area of a triangle T in the hyperbolic plane is simple. Suppose that three angles of a triangle have sizes (measured in radians) equal to α, β, and γ. Then the area[5] of T is

$$\pi - \alpha - \beta - \gamma. \tag{7.2}$$

Expression (7.2) is always equal to 0 for a triangle in the Euclidean plane, but is always positive for triangles in the hyperbolic plane. It is also known as the *angular defect* of the triangle.

One implication of this is that the area of a triangle cannot exceed π. If we allow the corners of a triangle to all be ideal points (as in Figure 7.6), then the lengths of the sides of the triangle are all infinite, but the area is $\pi - 0 - 0 - 0 = \pi$, which is finite!

---

[5]For the experts: This formula depends on choosing the scaling constant for the hyperbolic plane, K, to be equal to 1.

For your consideration

There are no squares in the hyperbolic plane. The nearest analog is a regular quadrilateral: one whose four angles are all the same and whose sides are all the same. In that case, the four angles are all less than $\pi/2$ radians.

Suppose that the angles of an n-gon in the hyperbolic plane have sizes $\theta_1, \theta_2, \ldots, \theta_n$. How large (and how small) can the sum of the angles be? Given the angles, what is the area of the n-gon? In Euclidean geometry, can you determine the area of a polygon just from its angles?

Part III: Infinite Things

Chapter 8

Counting Infinity

> Some infinities are bigger than other infinities.
> —John Green, *The Fault in Our Stars*

What can possibly be bigger than infinity? Is it not the largest number imaginable? Amazingly, when we think of numbers as a means to count collections, we find that, indeed, some infinities are larger than others!

8.1 Counting things

Numbers count: Numbers and counting are the conceptual origins of mathematics. The number 8 has a clear meaning of "eight things." Not all numbers are answers to counting problems; we can't have -3.5 things in our grocery basket. Still, just like 8, is there a way to think of ∞ as an answer to a counting problem?

The answer is certainly yes. For example, the set of all natural numbers

$$\mathbb{N} = \{0, 1, 2, 3, \ldots\}$$

has infinitely many members. Here are some other sets that contain infinitely many members:

- The integers, $\mathbb{Z} = \{\ldots, -3, -2, -1, 0, 1, 2, 3, \ldots\}$.

- The rational numbers, $\mathbb{Q}$. This is the set of all numbers that can be expressed as a fraction a/b where a and b are integers (and $b \neq 0$).

- The real numbers, $\mathbb{R}$. This is the set of all numbers that can be represented by decimals. It includes all the rational numbers, but also irrational numbers such as $\sqrt{2}$ and π.

These four sets are nested. The natural numbers are a subset of the integers, the integers are a subset of the rational numbers, and the rational numbers are a subset of the real numbers. In notation:

$$\mathbb{N} \subseteq \mathbb{Z} \subseteq \mathbb{Q} \subseteq \mathbb{R}.$$

All four contain infinitely many members. Does that mean that each has the same number of elements as the next? The answers are yes, yes, and no. Let's see why.

8.2 Sets, functions, bijections

While counting may be the conceptual start to mathematics, the actual foundation of mathematics rests on the concepts of *set* and *function*. Let's review those.

Sets

A *set* is an unordered, repetition-free collection of things. Small sets can typically be specified by listing elements between curly braces. For example,

$$A = \{1, 3, 8\}$$

is a set that contains three members: the numbers 1, 3, and 8.

Sets are *unordered*. This means that it doesn't matter how we list the elements; all that matters is which things are members and which are not. All of the following are completely equivalent ways to express the set A:

$$A = \{1, 3, 8\} = \{8, 3, 1\} = \{1, 8, 3\} = \{3, 1, 8\} \text{ and so on.}$$

Sets are *repetition-free*. This means that an object either is or is not a member of a set. An object cannot be in a set "more than once." If we write $\{1, 1, 3, 8\}$, the repetition of the number 1 does not means that 1 is a "double" member of the set. The set $\{1, 1, 3, 8\}$ is exactly the same as $\{1, 3, 8\}$. Both sets have precisely three elements: the numbers 1, 3, and 8.

Here are a few important notations for dealing with sets:

- $a \in A$. The symbol $\in$ stands for *is a member of* or *is an element of*. The meaning is that object a is a member of the set A. For example, if $A = \{1, 3, 8\}$, then $1 \in A$ is a true statement, but $2 \in A$ is a false statement.

- $A \subseteq B$. The symbol $\subseteq$ stands for *is a subset of*. When we write $A \subseteq B$, we are asserting that every element of A is also an element of B. It is possible that A and B are the same set, or that B contains all the elements of A and more. For example, $\{1, 3, 8\} \subseteq \{1, 2, 3, 4, 8\}$ is true, but $\{1, 2, 3\} \subseteq \{1, 3, 8\}$ is false.

- $|A|$. The notation $|A|$ stands for the *size* of A; that is, $|A|$ is the number of elements in the set A. For example, if $A = \{1, 3, 8\}$, then $|A| = 3$. For the moment, we only use this notation for finite sets, but we will expand to infinite sets subsequently. The size of a set is also known as its *cardinality*.

Functions

A typical first encounter with the concept of function is often in the context of an algebraic expression. For example, the function $f(x) = x^2 + 1$ takes a number as its input, squares it, adds one, and then returns the result as its output.

Functions, however, need not be specified by algebraic formulas, nor do the inputs and outputs have to be numbers. Schematically, a function looks like this:

$$x \longrightarrow \boxed{f} \longrightarrow y$$

A function is, loosely speaking, a rule that takes an input and returns an output.

The function $f(x) = x^2 + 1$ takes numbers as input and returns numbers as output. For example, $f(3) = 10$. It does not make sense to apply this function to a triangle.

When we specify a function, we need to be explicit about the set of allowable inputs (the domain of the function) and the set of possible outputs. To this end, we have the following notation:

$$f: A \to B. \tag{8.1}$$

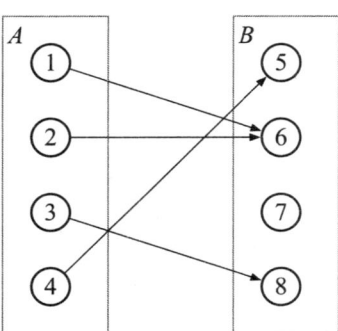

Figure 8.1: Visualization of the function $f: \{1,2,3,4\} \to \{5,6,7,8\}$ defined in (8.2). There is an arrow from each input value in A to its corresponding output value. For example, there is an arrow from 3 to 8 because $f(3) = 8$.

In this notation, f is a function, A is the set of allowable inputs, and B is a set that includes all of the possible outputs. To be explicit, the notation (8.1) means this:

- The function f is applicable to every element of A.

- Given any element a in A, the result $f(a)$ must be an element of the set B.

For example, suppose $A = \{1,2,3,4\}$ and $B = \{5,6,7,8\}$. We can define a function from A to B by specifying for each $a \in A$ the value that f takes in B, like this:

$$f(1) = 6, \quad f(2) = 6, \quad f(3) = 8, \quad \text{and} \quad f(4) = 5. \qquad (8.2)$$

Notice that element 7 in the set B is never the output of the function f. That's acceptable. All we are requiring is that every element of A is transformed into some element of B by the function f. Figure 8.1 provides a way to visualize this function.

Bijections

Given two sets, A and B, how might we decide if they have the same number of elements? One way is to see if the numbers $|A|$ and $|B|$ are equal; that is, we count how many elements are in each of the sets and check if those counts agree. This make sense if the sets are finite, in which case $|A|$ and $|B|$ are natural numbers.

An alternative method is to find an exact matching between the two sets. This is the idea behind the notion of a *bijection*.

Let A and B be sets, and let f be a function from A to B; that is, $f: A \to B$. Our aim is to develop language to express the idea that f gives an exact matching between the elements of the set A and those in the set B.

To this end, we ask that f be *one-to-one* and *onto*, concepts that we explain now.

- A function $f: A \to B$ is called *one-to-one* provided no two inputs give the same output. In other words, if a and a' are different members of the input set, A, then we must have $f(a) \neq f(a')$.

  The function specified in (8.2) (and illustrated in Figure 8.1) is not one-to-one because there are two different elements of the set A that yield the same output: both $f(1) = 6$ and $f(2) = 6$. Visually, the function is not one-to-one because there are two arrows pointing into 6. In order to be one-to-one, there can be at most one arrow pointing into each element of B.

  On the other hand, the function $g: \mathbb{R} \to \mathbb{R}$ defined by $g(x) = 2^x$ is one-to-one. If s and t are different real numbers, then $g(s) \neq g(t)$, as visualized in Figure 8.2.

- A function $f: A \to B$ is called *onto* provided every element of B is the output of f. Stated differently, for every $b \in B$ there must be an $a \in A$ such that $f(a) = b$.

  The function specified in (8.2) (and illustrated in Figure 8.1) is not onto. There is no element $a \in A$ with $f(a) = 7$. Visually, the function is not onto because there is no arrow pointing into 7. In order for a function to be onto, every element of B should have at least one arrow pointing to it.

  The function $g: \mathbb{R} \to \mathbb{R}$ defined by $g(x) = 2^x$ is also not onto because the output of g is always positive.

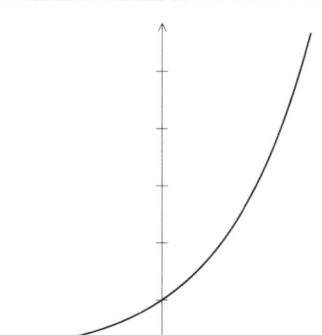

Figure 8.2: Graph of the function $g: \mathbb{R} \to \mathbb{R}$ defined by $g(x) = 2^x$. Notice that if $s \neq t$, then $g(s) \neq g(t)$. This can be visualized by noticing that a horizontal line in the diagram intersects the graph of the function in at most one point.

The function g is *not* onto because $g(x)$ is never zero or negative.

However, the function $h: \mathbb{R} \to \mathbb{R}$ defined by $h(x) = x^3 - 4x$ is onto. Examine Figure 8.3 and notice that all horizontal lines intersect the graph of this function. This means that for any y value there is at least one value x with $h(x) = y$.

For a function between finite sets, the properties of being one-to-one and onto have implications about the sizes of the sets.

Let $f: A \to B$ for finite sets A and B.

- If $|A| > |B|$, then A has more elements than B. In this case, it is impossible for f to be one-to-one; there are simply not enough elements in B to receive single elements of A.[1]

  It follows that if $f: A \to B$ is one-to-one, then $|A| \leq |B|$.

---

[1] This is a formal statement of what mathematicians call the *pigeon-hole principle*: If n pigeons want to roost in m pigeon holes, and if $n > m$, then some pigeons have to share living quarters.

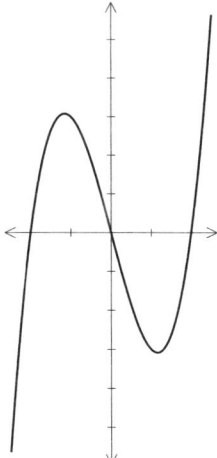

Figure 8.3: Graph of the function $h: \mathbb{R} \to \mathbb{R}$ defined by $h(x) = x^3 - 4x$. This function is onto because its outputs include every possible real number. This can be visualized by noticing that every horizontal line intersects the graph.

The function h is *not* one-to-one because (for example) $h(-2) = h(2)$.

- On the other hand, if $|A| < |B|$, then $f: A \to B$ cannot be onto. There simply aren't enough elements in A so that each element of B can be an output of the function.

It follows that if $f: A \to B$ is onto, then $|A| \geq |B|$.

When a function $f: A \to B$ is both one-to-one and onto, we call f a *bijection*.[2]

We conclude that for finite sets A and B, if $f: A \to B$ is a bijection, then $|A| = |B|$. The converse is also true: If A and B have finitely many elements, then there is a bijection from A to B.

---

[2]The word *bijection* has an interesting etymology. Using language that has mostly faded from mathematical usage, a one-to-one function is called an *injection* and an onto function is called a *surjection*. A function that is both kinds of "-jection" is therefore called a *bijection*.

> **Comparing Finite Sets A and B**
>
> - There is a one-to-one function $f: A \to B$ if and only if $|A| \le |B|$.
> - There is an onto function $f: A \to B$ if and only if $|A| \ge |B|$.
> - There is a bijection $f: A \to B$ if and only if $|A| = |B|$.

8.3 Finite and infinite sets

What does it mean for a set to be finite? ... to be infinite?

The intuitive idea is that a finite set "stops" and an infinite set "goes on forever." However, these descriptions are imprecise. We can use the language of functions to do better. We give two answers.

Our first answer relies on using the natural numbers, $\mathbb{N} = \{0, 1, 2, \ldots\}$. If A is a finite set, then the number of elements in that set, $|A|$, is a natural number.

A set A is *finite* if there is a bijection between A and a set of the form $\{1, 2, \ldots, n\}$ where n is a natural number. (We consider the empty set, $\emptyset$, the set with no elements, to be finite. When $n = 0$, the notation $\{1, 2, \ldots, n\}$ reduces to the empty set.)

If there is no such bijection of the form $f: A \to \{1, 2, \ldots, n\}$, then we say that A is infinite.

There is an alternative way to define finite and infinite sets that does not rely on the natural numbers. The rationale for having an alternative definition is that one may think of the natural numbers as being properties of finite sets. The sets $\{a, b\}$ and $\{x, y\}$ both have the property of "twoness." Hence it is desirable to have notions of finite and infinite that do not rely on presupposing numbers.

The idea is to consider one-to-one functions from a set to itself, $f: A \to A$. For example, suppose A is the set $\{a, b, c, d\}$ where the letters $a, b, c,$ and d stand for different objects. Let's construct some one-to-one functions from A to itself.

The easiest such one-to-one function has each element of A associated with itself:
$$f(a) = a, \quad f(b) = b, \quad f(c) = c, \quad \text{and} \quad f(d) = d.$$
However, there are other possible one-to-one functions, such as this:
$$f(a) = b, \quad f(b) = a, \quad f(c) = d, \quad \text{and} \quad f(d) = c.$$
In both cases, the one-to-one function is also onto; they are bijections.

Is there a one-to-one function from $A = \{a, b, c, d\}$ to itself that is *not* onto? Clearly, no! In order for the function to be one-to-one, once we assign a value to $f(a)$, we cannot reuse that value for any other element of A. Therefore, $f(b)$ needs to be something different than $f(a)$. Likewise, $f(c)$ is something else again, leaving only a single possibility for $f(d)$.

For the set $A = \{a, b, c, d\}$, all one-to-one functions $f: A \to A$ must also be onto.

Let's consider the same issue with the natural numbers, $\mathbb{N}$. Is a one-to-one function $f: \mathbb{N} \to \mathbb{N}$ necessarily onto? For example, we might have this function:
$$f(0) = 0, \quad f(1) = 1, \quad f(2) = 2, \quad f(3) = 3, \quad \text{and so on.}$$
Expressing f by a formula is easy: $f(x) = x$. This function is onto. But the question we are asking is this: Are all one-to-one functions $f: \mathbb{N} \to \mathbb{N}$ onto? The answer is no, and here is a simple example:
$$f(0) = 1, \quad f(1) = 2, \quad f(2) = 3, \quad f(3) = 4, \quad \text{and so on.}$$
Writing this as a formula, $f(x) = x + 1$. This function is one-to-one, but it is *not* onto because there is no $x \in \mathbb{N}$ with $f(x) = 0$.

This dichotomy (whether a one-to-one function is necessarily onto, versus is not necessarily onto) provides a clear basis to distinguish finite from infinite sets. Summarizing:

Finite and Infinite Sets

- If every one-to-one function $f: A \to A$ is also onto, then the set A is *finite*.

- If there is a one-to-one function $f: A \to A$ that is not onto, then A is *infinite*.

Here are some additional examples.

- The integers, $\mathbb{Z}$, form an infinite set. The function $f: \mathbb{Z} \to \mathbb{Z}$ defined by $f(x) = 2x$ is a one-to-one function. (If $x \neq y$, then certainly $2x \neq 2y$, i.e., $f(x) \neq f(y)$.) This function is not onto because there is no integer x such that $f(x) = 1$ (because 1 is odd).

- The interval $[0, 1]$ is an infinite set.[3] The function $f: [0, 1] \to [0, 1]$ defined by $f(x) = x/2$ is one-to-one (if $x \neq y$, then $x/2 \neq y/2$). However, it is not onto because there is no $x \in [0, 1]$ with $f(x) = \frac{3}{4}$.

- The rational numbers, $\mathbb{Q}$, form an infinite set. The function $f: \mathbb{Q} \to \mathbb{Q}$ defined by $f(x) = x^3$ is a one-to-one function. (If $a^3 = b^3$, we must have $a = b$.) However, it is not onto. There is no rational number x with $f(x) = x^3 = 2$ because $\sqrt[3]{2}$ is irrational.

Hilbert's hotel

There is an amusing narrative ascribed to the nineteenth/twentieth-century German mathematician David Hilbert.[4] The story is based on one-to-one functions from $\mathbb{N}$ to itself that are not onto, specifically the functions $f(x) = x + 1$ and $g(x) = 2x$.

Imagine a hotel with infinitely many rooms labeled using the natural numbers: 0, 1, 2, 3, and so forth. You arrive at the hotel—but, alas, every room is occupied.

Were this an ordinary hotel, you would be out of luck. Fortunately, there is an easy way to create a vacancy. Ask the person in room 0 to move to room 1, the person in room 1 to move to room 2, and so on. Symbolically, the occupant of room n moves to room $n + 1$. Now room 0 is empty, so you are able to check in!

There's more to the story. It turns out that Hilbert owns *two* such hotels; let's call them Hotel A and Hotel B. One evening there is a power failure at Hotel B. The heat won't come on, and the water won't run. We need to find a place for all the occupants in Hotel B. The previous trick of asking everyone to move to the next room won't work, no matter how many times we try it.

---

[3] The notation $[a, b]$ stands for the set of all real numbers x with $a \leq x \leq b$.

[4] A delightful, animated illustration of the Hilbert hotel is presented in the Netflix film *A Trip to Infinity*.

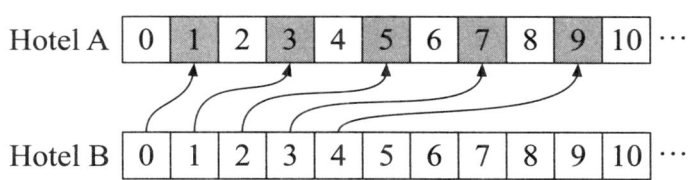

Figure 8.4: After the occupants in Hotel A move from room n to room $2n$, there are enough unoccupied rooms to accommodate all the residents from Hotel B.

Instead, we ask every occupant of Hotel A to move to the room whose number is double their currently assigned number. That is, the occupant of room n moves to room $2n$.

After this, the people in Hotel B can move into the odd-numbered rooms. The person in room 0 moves into room 1, the person in room 1 moves into room 3, the person in room 2 moves into room 5, and so on. In general, the occupant in Hotel B's room k moves into the empty room $2k + 1$ in Hotel A (Figure 8.4).

8.4 Comparing infinite sets

We have used functions to nail down two concepts:

- Is a set A infinite? Yes, when there is a function from A to itself that is one-to-one but not onto.

- Are sets A and B the same size? Yes, when there is a bijection from A to B.

We now consider the question: Are all infinite sets the same size?

Let's start by comparing the natural numbers, $\mathbb{N}$, and the integers, $\mathbb{Z}$. At first blush, it might seem that there are twice as many integers as natural numbers because the integers include all the natural numbers, and then another infinity of negative integers.

However, we assert that the two sets have the same size by providing a bijection $f: \mathbb{N} \to \mathbb{Z}$. We present the bijection as a table:

x	0	1	2	3	4	5	6	7	8	...
	↓	↓	↓	↓	↓	↓	↓	↓	↓	
$f(x)$	0	1	−1	2	−2	3	−3	4	−4	...

Examining this table, we see that every natural number (listed in the top row) is paired one-for-one with every integer (in the bottom row); this is a bijection from $\mathbb{N}$ to $\mathbb{Z}$, and therefore these sets have the same size.

Though not necessary, this function may be expressed using a formula like this:

$$f(n) = \begin{cases} (n+1)/2 & \text{if } n \text{ is odd, and} \\ -n/2 & \text{if } n \text{ is even.} \end{cases}$$

Perhaps this is not surprising. Both sets are infinite, so one might expect that they have the same "size."

However, we shall now present infinite sets that are *not* the same "size."

Power sets

Let A be a set. The *power set* of A is a new set whose elements are all the subsets of A. The notation for the power set of A is $\mathscr{P}(A)$.

Let's look at an example. Suppose A is the set $\{1, 2, 3\}$. The power set of A contains all possible subsets of A. That is,

$$\mathscr{P}(A) = \Big\{\varnothing, \{1\}, \{2\}, \{3\}, \{1,2\}, \{1,3\}, \{2,3\}, \{1,2,3\}\Big\}.$$

All told, $\mathscr{P}(A)$ contains 8 elements, each of which is a subset of A.

We show that it is impossible to have a bijection between a set A and its power set $\mathscr{P}(A)$ using a technique developed by the German-Russian nineteenth/twentieth-century mathematician Georg Cantor.

The basic idea is this: Suppose we are given any function $f: A \to \mathscr{P}(A)$. We show that f is not onto by constructing a subset X of A that is not the output of f.

Here is how we construct X. For an element a of A, the output of $f(a)$ is a subset of A. Since $f(a)$ is itself a set, it is meaningful to ask, is $a \in f(a)$?

For example: Let $A = \{1, 2, 3\}$. Perhaps $f(1) = \{1, 2\}$. In this case, $1 \in f(1)$. But it might also be the case that $f(1) = \{2, 3\}$, implying that $1 \notin f(1)$.

Returning to our discussion about a general $f: A \to \mathscr{P}(A)$, we define X to be the set of those elements x in A with the property that $x \notin f(x)$. In set-theory notation:

$$X = \{x \in A : x \notin f(x)\}.$$

We assert that there is no element a of A with $f(a) = X$. Here's why: We show that assuming $f(a) = X$ always leads to a contradiction.

There are two possibilities, $a \in f(a)$ or $a \notin f(a)$. Let's consider each possibility:

- If $a \in f(a)$, this means that $a \notin X$ because X is defined as the set of elements with the property $a \notin f(a)$. But $X = f(a)$, so we have both $a \in f(a) = X$ and $a \notin f(a) = X$. That's not possible!

- Otherwise, $a \notin f(a)$. By the definition of the set X, we know that $a \in X$. But, as before, $f(a) = X$, so we have determined that $a \notin X$ and $a \in X$. That's also not possible.

In other words, the assumption that there is an a such that $f(a) = X$ leads to the impossible conclusion that both $a \in X$ and $a \notin X$.

Therefore, there can be no bijection $f: A \to \mathscr{P}(A)$ because no such function can be onto, as explained in "No onto function from a set to its power set" on the next page.

Let's apply what we have learned to the set of natural numbers, $\mathbb{N}$. By Cantor's argument, there cannot be a bijection from $\mathbb{N}$ to $\mathscr{P}(\mathbb{N})$. However, there is a one-to-one (but not onto) function f from $\mathbb{N}$ to $\mathscr{P}(\mathbb{N})$ like this:

n	0	1	2	3	4	5	6	7	$\cdots$
	$\downarrow$	$\downarrow$	$\downarrow$	$\downarrow$	$\downarrow$	$\downarrow$	$\downarrow$	$\downarrow$	
$f(n)$	$\{0\}$	$\{1\}$	$\{2\}$	$\{3\}$	$\{4\}$	$\{5\}$	$\{6\}$	$\{7\}$	$\cdots$

As we emphasized in "Comparing finite sets A and B" on page 110, because there is no onto function from $\mathbb{N}$ to $\mathscr{P}(\mathbb{N})$, but there is a one-to-one function, the inexorable conclusion is that $\mathbb{N}$, although infinite, is smaller than $\mathscr{P}(\mathbb{N})$.

> **No onto function from a set to its power set**
>
> It is impossible to have an onto function f from a set A to its power set $\mathscr{P}(A)$. We illustrate the construction of the set X in the proof with a small example.
>
> Suppose $A = \{1, 2, 3, 4\}$ and define $f: A \to \mathscr{P}(A)$ as follows:
>
> $$f(1) = \{2, 3, 4\}, \quad f(2) = \{1, 2\},$$
> $$f(3) = \varnothing, \text{ and} \quad f(4) = \{1, 3, 4\}.$$
>
> The set X consists of those elements a for which $a \notin f(a)$. Checking each of the four elements:
>
> - $1 \notin f(1) = \{2, 3, 4\}$, therefore $1 \in X$.
> - $2 \in f(2) = \{1, 2\}$, therefore $2 \notin X$.
> - $3 \notin f(3) = \varnothing$, therefore $3 \in X$.
> - $4 \in f(4) = \{1, 3, 4\}$, therefore $4 \notin X$.
>
> Hence the set X is $\{1, 3\}$. Notice there is no element a in A with $f(a) = X$.

Repeating Cantor's argument for $\mathbb{N}$ versus $\mathbb{R}$

There is a variant of this same argument that shows that the natural numbers, $\mathbb{N}$, and the real numbers, $\mathbb{R}$, are not the same size. In particular, the interval $[0, 10]$ is already "bigger" than the natural numbers.

The general idea is the same as before. Suppose there is a function $f: \mathbb{N} \to [0, 10]$. We show that this function is not onto. That is, there is a real number $x \in [0, 10]$ with the property that $f(n) \neq x$ for any natural number n.

Imagine we create a chart showing this function. It would look something like the table in Figure 8.5. We show that no matter how the function is created, there is a real number x that ought to appear in the right-hand column, but has been missed.

The idea is to look at the nth digit in row n. In row 0, that's the digit to the left of the decimal point. In row 1, it's the first digit to the right of the

n	$f(n)$
0	4.5124589103...
1	0.2335905794...
2	2.1194097851...
3	7.2293591854...
4	6.5912401538...
5	0.8155488921...
6	1.5523785215...
⋮	⋮

Figure 8.5: A hypothetical bijection between the natural numbers and the real numbers in the interval $[0, 10]$.

decimal point, and so forth. This is indicated in the table by underlining the digit of interest.

We are ready to create a number x that is not on the right side of the table. The digits of x are determined by the underlined digits in the chart. The first step is to create a number by reading off the underlined digits. This gives the following number:

$$4.219248....$$

The second step is to "roll" the digits up by one. That is, each digit in this number is increased by 1, except 9 cycles back to 0. The result is the number we name x:

$$5.320359....$$

We now observe that x cannot possibly be on the right-hand side of the table. Can x be in row n of the table? Not possible because the nth digit of x is different from the nth digit of that number!

Hence the function $f: \mathbb{N} \to [0, 10]$ cannot be onto. Therefore, there is no bijection from $\mathbb{N}$ to $[0, 10]$. Please note that there are one-to-one functions $g: \mathbb{N} \to [0, 10]$ such as $g(n) = 10^{-n}$. We see that the interval $[0, 10]$ is "bigger" than the natural numbers $\mathbb{N}$. And since $[0, 10] \subseteq \mathbb{R}$, we have shown that $\mathbb{R}$ is "more infinite" than $\mathbb{N}$.

8.5 Transfinite cardinal numbers

When sets are finite, comparing sizes is simple. Let A and B be finite sets with $|A| = a$ and $|B| = b$ where a and b are natural numbers. There are three possibilities:

- A and B have the same size. In this case, $a = b$ and there is a bijection between A and B.

- A is smaller than B. In this case, $a < b$ and there is a one-on-one function $f: A \to B$, but no onto function.

- A is larger than B. In this case, $a > b$ and there is an onto function $f: A \to B$, but no one-to-one function.

Similar statements can be made for infinite sets, except we can't ascribe natural numbers to their sizes. In particular, for infinite sets A and B, the following is known (though the proofs are technical and beyond our scope):

- There is always either a one-to-one function $f: A \to B$ or a one-to-one function $f: B \to A$, or both.

- There is a one-to-one function $f: A \to B$ if and only if there is an onto function $g: B \to A$.

- If there are one-to-one functions $f: A \to B$ and $g: B \to A$, then there is a bijection between A and B. (This is known as the Schröder-Bernstein or Cantor-Schröder-Bernstein theorem.)

The upshot is that any two infinite sets can be compared. One of the following must be true:

- A and B have the same size, in which case there is a bijection $f: A \to B$.

- A is smaller than B, in which case there is a one-to-one function $f: A \to B$ but no possible onto function.

- A is larger than B, in which case there is an onto function $f: A \to B$, but no possible one-to-one function.

What we see is that the situation is exactly the same as for finite sets, except we don't have numbers to assign to the sizes of the sets. It is inadequate to write $|A| = \infty$ and $|B| = \infty$ because this doesn't provide enough information to say whether they have the same size (there is a bijection between them) or if one is larger than the other.

To remedy this situation, Georg Cantor invented a new category of numbers that are called *transfinite cardinal numbers*. The sizes of finite sets are natural numbers; these new numbers specify the sizes of infinite sets.

The natural numbers, $\mathbb{N}$, are the smallest-size infinite set. Cantor used the notation[5] $\aleph_0$ to stand for the number of elements $\mathbb{N}$. In notation $|\mathbb{N}| = \aleph_0$.

It might "feel" as though there are twice as many integers as natural numbers; as we saw in Section 8.4, there is a bijection between $\mathbb{N}$ and $\mathbb{Z}$. This means that $\mathbb{Z}$ has the same size as $\mathbb{N}$; namely, $|\mathbb{Z}| = \aleph_0$.

Sets of size $\aleph_0$ are called *countable* because they can be put in one-to-one correspondence with the positive integers.

What about the rational numbers, $\mathbb{Q}$? It might "feel" as though there are a lot more rational numbers than integers, but in fact the size of $\mathbb{Q}$ is also $\aleph_0$. Let's see why.

First, we write rational numbers in reduced form. That is, we cancel any common factors between numerator and denominator, rendering $\frac{8}{12}$ as $\frac{2}{3}$.

Then we list the positive rational numbers in groups. The rational number $\frac{a}{b}$ is in group $a + b$. Within groups, we list the numbers by numerator. For example, in group 10 we have these numbers in this order:

$$\frac{1}{9} \quad \frac{3}{7} \quad \frac{7}{3} \quad \frac{9}{1}$$

but we don't have $\frac{2}{8}$ because $\frac{2}{8} = \frac{1}{4}$, and $\frac{1}{4}$ is in group 5.

By this method, here is the beginning of the list of positive rational numbers marked by their group number:

$$\underbrace{\frac{1}{1}}_{2} \quad \underbrace{\frac{1}{2} \frac{2}{1}}_{3} \quad \underbrace{\frac{1}{3} \frac{3}{1}}_{4} \quad \underbrace{\frac{1}{4} \frac{2}{3} \frac{3}{2} \frac{4}{1}}_{5} \quad \underbrace{\frac{1}{5} \frac{5}{1}}_{6} \quad \underbrace{\frac{1}{6} \frac{2}{5} \frac{3}{4} \frac{4}{3} \frac{5}{2} \frac{6}{1}}_{7} \quad \underbrace{\frac{1}{7} \frac{3}{5} \frac{5}{3} \frac{7}{1}}_{8} \quad \underbrace{\frac{1}{8} \frac{2}{7} \frac{4}{5} \frac{5}{4} \frac{7}{2} \frac{8}{1}}_{9} \quad \underbrace{\frac{1}{9} \frac{3}{7} \frac{7}{3} \frac{9}{1}}_{10} \cdots$$

---

[5]The symbol $\aleph$ is the Hebrew letter aleph.

With this, we can create a bijection between $\mathbb{Z}$ and $\mathbb{Q}$ by listing the negative rationals in the reverse order, then zero, and then the positive rational numbers. The end result looks like this:

-7	-6	-5	-4	-3	-2	-1	0	1	2	3	4	5	6	7
↓	↓	↓	↓	↓	↓	↓	↓	↓	↓	↓	↓	↓	↓	↓
$-\frac{2}{3}$	$-\frac{1}{4}$	$-\frac{3}{1}$	$-\frac{1}{3}$	$-\frac{2}{1}$	$-\frac{1}{2}$	$-\frac{1}{1}$	0	$\frac{1}{1}$	$\frac{1}{2}$	$\frac{2}{1}$	$\frac{1}{3}$	$\frac{3}{1}$	$\frac{1}{4}$	$\frac{2}{3}$

going off to infinity in both directions.

This gives a bijection between $\mathbb{Z}$ and $\mathbb{Q}$, and so $|\mathbb{Q}| = |\mathbb{Z}| = \aleph_0$. The rational numbers may look superficially "bigger" than the integers, but $\mathbb{Q}$, too, is a countable set.

On the other hand, the real numbers, $\mathbb{R}$, really are more numerous than the integers (and other sets of size $\aleph_0$). Because it is impossible to match up the reals with the positive integers, we say that $\mathbb{R}$ is *uncountable*.

The continuum hypothesis

We see that $\mathbb{N}$, $\mathbb{Z}$, and $\mathbb{Q}$ all have the same size: $\aleph_0$. However, as shown in Section 8.4, there is no bijection between $\mathbb{N}$ and $\mathbb{R}$. Since $\mathbb{N} \subseteq \mathbb{R}$, we see that $\mathbb{R}$ is bigger than $\mathbb{N}$; that is, $|\mathbb{R}| > \aleph_0$.

The transfinite cardinal number representing the size of the real numbers is $\mathfrak{c}$. (It is also the case that $|\mathscr{P}(\mathbb{N})| = |\mathbb{R}| = \mathfrak{c}$, but showing this is a bit technical and beyond our scope.) The real numbers are known as a *continuum*, and the symbol $\mathfrak{c}$ stands for *continuum*.[6]

We have $|\mathbb{N}| = \aleph_0$ and $|\mathbb{R}| = \mathfrak{c}$. Why the different types of notation?

Cantor defined $\aleph_1$ to be the next larger size of an infinite set above $\aleph_0$. That is, it is the size of a set A with $|A| > \aleph_0$, but for which there is no subset of A with size strictly between $|A|$ and $|\mathbb{N}|$.

It is natural to ask, is $\mathfrak{c} = \aleph_1$? Cantor believed the answer to this question is yes. In other words, he conjectured that the following is true:

Continuum Hypothesis
There is no set A with $\aleph_0 < |A| < \mathfrak{c}$.

In 1963, American mathematician Paul Cohen gave a most unsettling answer to the question: Is the continuum hypothesis true?

---
[6]The symbol $\mathfrak{c}$ is the Gothic letter c.

Cohen's answer: It depends. To be clear: The answer is not "We don't know." The answer is "It depends. Can be yes. Can be no." This is bizarre because one would expect that questions of this form should have a definitive answer, even if we don't know what that answer is.

Let's elaborate on Cohen's result. We have described *sets* as unordered, repetition-free collections of things. We have not given a mathematical, iron-clad definition but rather relied on the intuitive meaning of the word *collection*. The lack of a rigorous notion of *set* leads to contradictions (most notably, Russell's paradox, explained below).

To avoid such contradictions, specific axioms of set theory have been developed, with the system proposed by Ernst Zermelo and Abraham Fraenkel being the one most familiar to and widely adopted by mathematicians. The ZF axioms, as their formulation is called, are akin to Euclid's axioms for geometry.

In plane geometry, does the parallel postulate hold? It depends! It depends if we're talking about the Euclidean plane (in which case, yes) or either the hyperbolic or projective plane (in which case, no). See Chapters 6 and 7.

The situation here is analogous. What exactly do we mean by *set*? This turns out to be a subtle question, and, depending on one's notion of *set*, the continuum hypothesis may be true or it may be false.

Addendum: Russell's paradox

The following paradox, attributed to Bertrand Russell, underscores the need for a careful approach to set theory. The vague "collection" idea gets us in trouble, as the following example shows.

Recall that the notation $a \in A$ means that object a is a member of the set A. Also recall that sets themselves may be members of a set. This is the case when we consider power sets. For example, let $A = \{1, 2, 3\}$. Then $\mathscr{P}(A)$ is itself a set and $A \in \mathscr{P}(A)$.

This is how the Russell paradox arises. Typically, a set is not a member of itself.[7] We call a set *normal* if $A \notin A$. The set of real numbers, $\mathbb{R}$, is a normal set. All of its elements are numbers; we don't have $\mathbb{R} \in \mathbb{R}$.

Let X be the set of all normal sets. The question is: Is X normal?

If X is normal, it means that X should be included in the set of all normal sets—that is, that $X \in X$. But that means that X is not normal.

---

[7] A set is always a subset of itself. That's different.

On the other hand, if X is not normal, that means that $X \in X$, which in turn implies that X is in the set of all normal sets, implying that X is normal.

Summarizing:

- If X is normal, then X is not normal.

- If X is not normal, then X is normal.

And that's impossible!

This conundrum led to the careful, axiomatic approach to set theory.

Addendum: On beyond $\mathfrak{c}$

The set of real numbers, $\mathbb{R}$, and the set of natural numbers, $\mathbb{N}$, while both infinite, are not the same size. The size of $\mathbb{R}$ is larger than the size of $\mathbb{N}$. Is there a set larger than $\mathbb{R}$? Perhaps one might expect the set of complex numbers, $\mathbb{C}$, to be larger than $\mathbb{R}$. However, using an argument akin to that with which we showed that $\mathbb{Q}$ and $\mathbb{Z}$ have the same size, it is known that $|\mathbb{C}| = |\mathbb{R}| = \mathfrak{c}$.

So the question remains: Is there a set that has even more elements than the uncountable $\mathbb{R}$? The answer is yes, and we have already presented a way to create such a set: form the power set. As we have shown, there is never a bijection from a set A to its power set $\mathscr{P}(A)$. It follows that $|A| < |\mathscr{P}(A)|$ regardless of whether A is finite or infinite. Hence, $\mathscr{P}(\mathbb{R})$—the set of all subsets of $\mathbb{R}$—is of greater cardinality than $\mathbb{R}$ itself. In symbols:

$$\mathfrak{c} = |\mathbb{R}| < |\mathscr{P}(\mathbb{R})|.$$

There is an extension to the continuum hypothesis that says that there is no set whose size lies between that of $\mathbb{R}$ and that of $\mathscr{P}(\mathbb{R})$. Under the assumption of the so-called *generalized continuum hypothesis*, we have:

$$|\mathbb{N}| = \aleph_0, \quad |\mathbb{R}| = \aleph_1, \quad \text{and} \quad |\mathscr{P}(\mathbb{R})| = \aleph_2.$$

Continuing, $\aleph_3 = |\mathscr{P}(\mathscr{P}(\mathbb{R}))|$ and so on.

This gives us an infinite sequence of transfinite cardinals:

$$\aleph_0, \aleph_1, \aleph_2, \aleph_3, \ldots. \tag{8.3}$$

Are there transfinite cardinals larger than all of these?

There are. The next one after those listed in (8.3) is named $\aleph_\omega$, and here is how to create a set with that cardinality. For every natural number n, choose a set X_n of cardinality $\aleph_n$. Then let X be the union of all these sets; that is,

$$X = X_0 \cup X_1 \cup X_2 \cup \cdots.$$

The cardinality of X is necessarily greater than any $\aleph_n$ where $n \in \mathbb{N}$. What is that mysterious subscript ω? It is explained in Chapter 9.

8.6 Transcendental numbers

The fact that the real numbers are uncountably infinite leads to an elegant solution to the question: Are there any transcendental numbers? Let's see what is meant by *transcendental number* and how uncountable infinite sets lead to an affirmative answer.

We start by defining *algebraic numbers*. These are the roots of polynomials with integer coefficients.[8] For example, $\sqrt{2}$ is an algebraic number because it is a root of the polynomial $x^2 - 2$.

The algebraic numbers include all the rational numbers. The rational number a/b (where a and b are integers) is a root of the integer polynomial $bx - a$. The imaginary number i is algebraic because it is a root of the integer polynomial $x^2 + 1$.

Are all numbers algebraic?

Numbers (real or complex) that are not algebraic are called *transcendental*. Today, we know that numbers such as π are transcendental (this is not easy to prove), but in Cantor's day their existence was an open question. Here is an outline of his argument for the existence of transcendental numbers:

- Show that the set of algebraic numbers is countable; that is, it has size $\aleph_0$. (We'll explain why below.)

- Note that $\mathbb{R}$ has size $\mathfrak{c}$.

- Since $\mathfrak{c} > \aleph_0$, there are more—many more!—real numbers than algebraic numbers; we thus learn that "most" real numbers are transcendental.

---

[8] For simplicity, we call polynomials with integer coefficients *integer polynomials*.

A remarkable feature of Cantor's argument is that it firmly establishes the existence of transcendental numbers without ever giving an explicit example (such as π or e).

Here is a sketch of why the set of algebraic numbers is countable.

We start by defining the *height* of an integer polynomial. Suppose p is the integer polynomial

$$a_d x^d + a_{d-1} x^{d-1} + \cdots + a_1 x + a_0.$$

The height of p is the sum of its degree and the absolute values of its coefficients—that is,

$$d + |a_d| + |a_{d-1}| + \cdots + |a_1| + |a_0|.$$

For example, if $p = x^3 + 5x^2 - 4$, then the height of p is

$$3 + 1 + 5 + 0 + 4 = 13.$$

In this sum, the 3 is the degree of the polynomial and the other summands (1, 5, 0, and 4) are the absolute values of the coefficients.

For a given h, there can be at most finitely many integer polynomials with height h. This is because the number of terms is bounded (can't be more than h) and the coefficients are bounded (must lie in the range from $-h$ to h).

Let A_h be the set of all roots of integer polynomials with height h. Since a polynomial of degree d has (at most) d roots, and since there are finitely many integer polynomials of height h, the set A_h is finite.

Let A be the set of all algebraic numbers. Since every algebraic number is the root of an integer polynomial, we can write:

$$A = A_1 \cup A_2 \cup A_3 \cup A_4 \cup \cdots.$$

In other words, A is the union of countably many finite sets. With a bit of work (which we are omitting), one can show that the union of countably many finite sets is countable, and therefore $|A| = \aleph_0$.

Finally, since $|\mathbb{R}| = \mathfrak{c} > \aleph_0$, we see that there are more—infinitely more!—real numbers than algebraic numbers. Therefore, some—indeed, most!—real numbers are transcendental.

For your consideration

The set of natural numbers, $\mathbb{N}$, is countable; that is, $|\mathbb{N}| = \aleph_0$. However, the set of all subsets of $\mathbb{N}$ is uncountable; that is, $|\mathscr{P}(\mathbb{N})| > \aleph_0$. One way to express the uncountability of $\mathscr{P}(\mathbb{N})$ is that it is impossible to make an ordered list of the subsets of $\mathbb{N}$ that is all-inclusive.

However, the set of all *finite* subsets of $\mathbb{N}$ is countable. It is possible to make a list that contains all the finite subsets of $\mathbb{N}$.

Here is a *bad* way to start:

$$\{0\}$$
$$\{1\}$$
$$\{2\}$$
$$\{3\}$$
$$\vdots$$

This list starts with all the singleton subsets, and there's no possibility of getting to other finite subsets. Find a way to construct a list that is guaranteed to include all the finite subsets of $\mathbb{N}$.

Chapter 9

Order Infinity

> It turns out that an eerie type of chaos can lurk just behind a facade of order—and yet, deep inside the chaos lurks an even eerier type of order.
>
> —Douglas Hofstadter

Chapter 8 introduced new numbers to represent the sizes of infinite sets. These transfinite cardinal numbers include $\aleph_0$ and $\mathfrak{c}$ that stand for the sizes of $\mathbb{N}$ and $\mathbb{R}$, respectively. The fact that both $\mathbb{N}$ and $\mathbb{Z}$ have size $\aleph_0$ is a reflection of the fact that there is a bijection $f: \mathbb{N} \to \mathbb{Z}$.

Elements of a set are unordered. However, when imbued with an order, a different suite of transfinite numbers emerges that are even eerier than the $\aleph$ numbers.

9.1 Ordered sets

The sets $\{1, 2, 3\}$, $\{2, 3, 1\}$, and $\{3, 2, 1\}$ are all the same. The order in which the elements are listed is irrelevant.

In this chapter we enhance the idea of set; we consider *ordered sets*. The sets we consider come with an ordering relation for their elements. By this we mean that if A is an ordered set, then not only does it have elements, but it also has a binary relation $\prec$. The curved symbol $\prec$ is meant to look similar to, but not be identical with, the usual less-than symbol, $<$.

Here is an example: $\{1 \prec 5 \prec 2\}$. This is a three-element set that is ordered with 1 as its first element, 5 as its second element, and 2 as its third (and final) element. The ordered set $\{2 \prec 5 \prec 1\}$ has the same elements,

> **Order relations**
>
> For a binary relation $\prec$ to be considered an *order relation* for a set A, it needs to satisfy these criteria:
>
> - *Trichotomy property*: For any two elements a and b of the set A, exactly one of the following is true (and the other two are false): $a \prec b$, $b \prec a$, or $a = b$.
>
> - *Transitive property*: For any three elements a, b, and c of A, if $a \prec b$ and $b \prec c$, then $a \prec c$.

but their order is different. The notation $a \prec b$ means that a comes before b in the ordering. We can also say that a is less than b, even though $\prec$ might not be the usual less-than relation for real numbers. A formal definition of an ordering relation is presented in the boxed comment on this page.

From the perspective of counting, sets A and B have the same size provided there is a bijection $f: A \to B$. In this sense, the sets $\mathbb{N}$ and $\mathbb{Z}$ are alike; they both have the same number of elements: $\aleph_0$. However, as ordered sets (using their usual, $<$ order), they are quite different. One noticeable difference is that $\mathbb{N}$ contains a smallest (first) element, 0, whereas $\mathbb{Z}$ does not have a least element.

The set of rational numbers, $\mathbb{Q}$, also has $\aleph_0$ elements; it has the same size as $\mathbb{N}$ and $\mathbb{Z}$. However, as an ordered set, it is again quite different from both $\mathbb{N}$ and $\mathbb{Z}$. In either set, we note that if $a < b$, there are finitely many numbers x with $a < x < b$. Yet for the rational numbers, $\mathbb{Q}$, between any two elements there are infinitely many other rational numbers.

If we ignore ordering, then the three sets $\mathbb{N}$, $\mathbb{Z}$, and $\mathbb{Q}$ are alike, and bijections between them show how to relabel elements of one to correspond to the other. However, as ordered sets, they are distinct from each other.

In contrast, the natural numbers $\mathbb{N}$ and the positive integers $\mathbb{Z}^+ = \{1, 2, 3, \ldots\}$ are ordered sets that look exactly alike. Consider the bijection $f: \mathbb{N} \to \mathbb{Z}^+$ defined by $f(x) = x + 1$:

x	0	1	2	3	4	5	6	...
	↓	↓	↓	↓	↓	↓	↓	
$f(x)$	1	2	3	4	5	6	7	...

This is more than just a bijection from $\mathbb{N}$ to $\mathbb{Z}^+$. It is also *order preserving*, by which we mean that if a is before b in the first set, then $f(a)$ is before $f(b)$ in the second.

If A and B are ordered sets, we say they are *order isomorphic* provided there is an order-preserving bijection $f: A \to B$. In formal language:

- $f: A \to B$ is a bijection (is both one-to-one and onto), and

- if a and a' are elements of A with $a < a'$ in A, then $f(a) < f(a')$ in B.

In this language, $\mathbb{N}$ and $\mathbb{Z}^+$ are order isomorphic, but $\mathbb{Z}$ and $\mathbb{Q}$ are not.

Notice that if A and B are finite ordered sets, then they are order isomorphic if and only if they contain the same number of elements. For example, $A = \{1 < 3 < 2\}$ is order isomorphic to $B = \{9 < 8 < 7\}$, and the order-preserving bijection $f: A \to B$ is

$$f(1) = 9, \quad f(3) = 8, \quad \text{and} \quad f(2) = 7.$$

9.2 Well-ordered sets

One of the features that distinguishes $\mathbb{N}$ and $\mathbb{Z}$ as ordered sets is that $\mathbb{N}$ has a first element, 0, whereas $\mathbb{Z}$ has no first (least) element. Other ordered sets have a least element (such as the nonnegative rationals), but the natural numbers have an additional feature concerning least elements: they are well ordered.

We say that an ordered set A is *well ordered* provided every nonempty subset of A contains a least element.

The natural numbers, $\mathbb{N}$, are a well-ordered set, and here are some examples of subsets of the natural numbers and their least elements:

- Let O be the set of odd natural numbers. The least element of O is 1.

- Let T be the set of two-digit natural numbers. The least element of T is 10.

- Let P be the set of prime numbers. The least element of P is 2.

The positive integers, $\mathbb{Z}^+$, are also a well-ordered set. Indeed, any subset of a well-ordered set must also be well ordered. Further, any finite ordered set is well ordered.

By contrast, none of these sets are well ordered:

- The set of integers, $\mathbb{Z}$, is not well ordered. The negative integers are a subset of $\mathbb{Z}$ that does not contain a least element.

- The set of non-negative rational numbers, $\mathbb{Q}^{\geq 0}$, is not well ordered. Although this set has a least element, 0, it contains, as a subset, the positive rational numbers, $\mathbb{Q}^+$, which do not contain a least element.

- The unit interval $[0, 1]$ is not well ordered. This is the set of all real numbers x with $0 \leq x \leq 1$. Even though $[0, 1]$ contains a least element, 0, it is not well ordered. Consider the subset $A = (\frac{1}{3}, \frac{2}{3})$. This is the open interval containing those real numbers x with $\frac{1}{3} < x < \frac{2}{3}$. The set A does not contain a least element.

Combining ordered sets

We have two examples of well-ordered sets: finite ordered sets and the natural numbers, $\mathbb{N}$. The positive integers, $\mathbb{Z}^+$, are also a well-ordered set, but this set is not really different from $\mathbb{N}$ because $\mathbb{Z}^+$ and $\mathbb{N}$ are order isomorphic.

It is natural to ponder: *Are there other infinite well-ordered sets?*

What about the prime numbers?[1] They form a well-ordered set (by virtue of being a subset of $\mathbb{N}$), but they are order isomorphic to the natural numbers:

$\mathbb{N}$	0	1	2	3	4	5	6	7	...
	↓	↓	↓	↓	↓	↓	↓	↓	
P	2	3	5	7	11	13	17	19	...

Here is another well-ordered set that is truly different from finite sets and $\mathbb{N}$. Simply append to $\mathbb{N}$ one additional element (let's call it z) that comes after all the other natural numbers:

$$\{0 < 1 < 2 < 3 < \cdots < z\}.$$

---

[1] There are infinitely many primes, and proofs of this can be found in many books, including this author's *The Mathematics Lover's Companion*.

This new ordered set—let's call it, just for the moment, $\hat{\mathbb{N}}$—is not order isomorphic to any finite set (it has infinitely many elements) and is not order isomorphic to $\mathbb{N}$ (it has a largest element and $\mathbb{N}$ does not).

Further, $\hat{\mathbb{N}}$ is well ordered. Let A be any nonempty subset of $\hat{\mathbb{N}}$.

- If z is the only element of A, then it's the least element of A.

- Otherwise, the non-z elements of A are also elements of $\mathbb{N}$, and therefore there is among them a least element.

Here is another example that foreshadows a more general idea. Take the natural numbers, $\mathbb{N}$, followed by a full duplicate copy of the natural numbers, $\mathbb{N}'$. We call the new ordered set $\mathbb{N} \uplus \mathbb{N}$, and it looks like this:

$$\mathbb{N} \uplus \mathbb{N} = \{0 < 1 < 2 < 3 < \cdots < 0' < 1' < 2' < 3' < \cdots\}.$$

This is also a well-ordered set. Let A be any nonempty subset of $\mathbb{N} \uplus \mathbb{N}$.

- If A contains any elements from the "original" (lesser) copy of $\mathbb{N}$, then because $\mathbb{N}$ is well ordered, A has a least element.

- Otherwise, A contains only elements from the "cloned" copy of $\mathbb{N}$, and it follows that A has a least element, by virtue of $\mathbb{N}$ being well ordered.

In full generality, if A and B are ordered sets, we form a new ordered set $A \uplus B$ by listing all the elements of A followed by all the elements of B—that is, unless A and B are elements in common. In that case, simply take an order-isomorphic copy of B that is disjoint from A.

For example:

$$\{1 < 5 < 2\} \uplus \{7 < 1 < 9\} = \{1 < 5 < 2 < 7 < 1' < 9\}$$

where we replace 1 in the second set with a clone $1'$ to ensure that the two sets we are combining are disjoint.

The earlier example, $\hat{\mathbb{N}}$, can be expressed in this notation as $\mathbb{N} \uplus \{z\}$.

Products of ordered sets

The operation $\uplus$ acts like addition for ordered sets. We explore this explicitly in Section 9.3. Here we describe an operation that behaves like multiplication, and we use the times symbol, $\times$, to underscore that.

For any sets (ordered or otherwise) A and B, the set $A \times B$ is formed by creating all ordered pairs (a, b) where a is an element of A and b is an element of B. For example, the set $\{1, 2, 3\} \times \{4, 5\}$ contains all possible pairs (a, b) where a is one of 1, 2, or 3, and b is one of 4 or 5:

$$\{1, 2, 3\} \times \{4, 5\} = \{(1,4), (2,4), (3,4), (1,5), (2,5), (3,5)\}.$$

This operation, $\times$, is called the *Cartesian product* of the sets.

Suppose now that A and B are ordered sets; we wish to imbue $A \times B$ with an order as well. To this end, we use *lexicographic order*. This is completely analogous to alphabetical order. The word cat is alphabetically before dog because we compare the words by their first letter. If two words have the same first letter, we then compare using the second letter. This puts the word car before the word cry because a comes before r.

If A and B are ordered sets, we order the pair (a, b) before the pair (a', b') if either

- $a < a'$ or

- $a = a'$ and $b < b'$.

Stated differently, we order distinct pairs (a, b) and (a', b') by first comparing a and a'. If $a < a'$, then $(a, b) < (a', b')$, but if $a' < a$, then $(a', b') < (a, b)$. The remaining possibility is a tie in the first coordinate (i.e., $a = a'$); in this case we order the pairs based on the second coordinate. If $b < b'$, then $(a, b) < (a, b')$. Otherwise, if $b' < b$, we have $(a, b') < (a, b)$.

For example:

$$\{2 < 8 < 3\} \times \{2 < 5\} = \{(2,2) < (2,5) < (8,2) < (8,5) < (3,2) < (3,5)\}.$$

(Notice that we allow elements to appear in both A and B when we form their product $A \times B$; this is in contrast to $A \uplus B$ where we may have to replace B with an order-isomorphic copy that is disjoint from A.)

Let's consider $\mathbb{N} \times \mathbb{N}$. This is the set of all possible ordered pairs (a, b) where $a, b \in \mathbb{N}$ is ordered lexicographically. It looks like this:

$$\begin{aligned}\mathbb{N} \times \mathbb{N} = \{&(0,0) < (0,1) < (0,2) < (0,3) < \cdots < \\ &(1,0) < (1,1) < (1,2) < (1,3) < \cdots < \\ &(2,0) < (2,1) < (2,2) < (2,3) < \cdots \cdots\}.\end{aligned}$$

In this way, not only is $\mathbb{N} \times \mathbb{N}$ an ordered set, but it is also well ordered; here's why. Let X be any nonempty subset of $\mathbb{N} \times \mathbb{N}$. We need to explain why X has a least element.

Consider just the first coordinates of the pairs (a, b) in X. Let A be the set of first coordinates that appear in X. For example, if $(9, 17) \in X$, then $9 \in A$. Since A is a nonempty subset of $\mathbb{N}$, it has a least element; let's call it a.

Now let B be the set of all numbers y such that $(a, y) \in X$. That is, we look at elements of X with first coordinate a, and gather all the second coordinates of those pairs into B. Since B is a nonempty subset of $\mathbb{N}$, it, too, has a least element b.

Focus on the pair (a, b), which is a member of X. There is no ordered pair with smaller first coordinate, and, among those pairs with a as first coordinate, there is no pair in X with smaller second coordinate. Therefore, (a, b) is the least element of X.

This discussion shows that $\mathbb{N} \times \mathbb{N}$ is a well-ordered set, but the only information we used about $\mathbb{N}$ is that it is well ordered. Indeed, if A and B are any well-ordered sets, so is their Cartesian product $A \times B$.

9.3 Ordinal numbers

Let's review: The size of a finite set is given by a natural number, $\mathbb{N}$. If A and B are sets with, say, 5 elements each, then we know there is a bijection $f: A \to B$. Conversely, if A and B are finite sets and if there is bijection $f: A \to B$, then we know that A and B have the same number of elements.

This idea extends to infinite sets. If two sets A and B have the same transfinite cardinality, say $\aleph_0$, then we know there is a bijection $f: A \to B$. Again, the converse holds. If there is a bijection $f: A \to B$, then there is a transfinite cardinal that describes the number of elements in both sets.

In summary, cardinal numbers are labels that enable us to know if there is a bijection between two sets: There is a bijection if and only if both sets are labeled by the same cardinal number (either a natural number if finite or a transfinite cardinal if infinite).

We now introduce *ordinal numbers* as descriptors of well-ordered sets: When A and B are order-isomorphic well-ordered sets, they are to be assigned the same ordinal number.

> **Can all sets be well ordered?**
>
> The integers, ℤ, in their usual order do not form a well-ordered set. For example, the subset of negative integers does not have a least element.
>
> However, the integers can be reordered to give a well ordering. Here are two alternative arrangements of the integers that are well orderings:
>
> $$0 < -1 < 1 < 2 < -2 < 3 < -3 < \cdots$$
>
> and
>
> $$0 < 1 < 2 < 3 < \cdots < -1 < -2 < -3 < \cdots.$$
>
> Although a set A might not be presented with a well ordering, it may be possible to arrange the elements of A into a well-ordered set. This leads to the question: Can every set have its elements arranged into a well ordering?
>
> Just as in the case of the continuum hypothesis (see page 120), the answer is: It depends. The answer depends on one's notion of *set*. The standard ZF axioms are often supplemented by an additional requirement known as the Axiom of Choice, which is often abbreviated AC. It would take us too far afield to delve into the substance of the AC, but if one takes the ZF axioms together with AC, one arrives at a notion of *set* in which every set can be well ordered. On the other hand, if one's notion of *set* is such that AC does not hold, then there are sets that cannot be well ordered.

Let's begin with finite sets. Any two ordered sets with n elements are order isomorphic. For example, both of these finite ordered sets have 4 elements:

$$A = \{3 < 5 < 1 < 9\} \quad \text{and} \quad B = \{2 < 8 < 7 < 1\}$$

and here is an order-preserving bijection $f: A \to B$:

$$f(3) = 2, \quad f(5) = 8, \quad f(1) = 7, \quad f(9) = 1.$$

For *finite* ordered sets, the number of elements in a set is a perfectly good label to determine if two sets are order isomorphic. In other words, the

> **Non-mathematical use of the term *ordinal number***
>
> It is common for mathematicians to repurpose everyday words to describe mathematical concepts. When mathematicians say that multiplication *distributes* over addition, they mean that $x \cdot (y+z) = x \cdot y + x \cdot z$. The connection to doling out items to recipients is metaphorical at best.
>
> Such is also the case with the term *ordinal number*. In common usage, this refers to numbers in an alternative grammatical form that indicates position as opposed to quantity. For example, I might be the *third* person standing in a line of *ten* waiting to enter a restaurant. The word *ten* tells us the quantity of people and the word *third* tells my position.
>
> In mathematics, an *ordinal number* is a descriptor of a well-ordered set. Two well-ordered sets are order isomorphic if and only if they are assigned the same ordinal number. The mathematical usage has an entirely different meaning altogether.

natural numbers, $\mathbb{N}$, serve both as cardinal numbers (is there a bijection between finite unordered sets?) and as ordinal numbers (is there an order-preserving bijection between finite ordered sets?).

The situation changes when we consider infinite well-ordered sets. For example, consider these:

$$\mathbb{N} = \{0 < 1 < 2 < 3 < 4 < \cdots\} \quad \text{and}$$
$$\mathbb{N} \uplus \{z\} = \{0 < 1 < 2 < 3 < 4 < \cdots < z\}.$$

Ignoring the ordering, both sets contain $\aleph_0$ elements, but they are not order isomorphic. The cardinal number $\aleph_0$ is insufficient to distinguish these.

To remedy this, we create a new category of numbers: the *ordinal* numbers that describe well-ordered sets. For finite ordered sets, we simply use natural numbers.

The set of natural numbers, $\mathbb{N}$, is the simplest infinite well-ordered set. We introduce a new number,[2] ω, to describe $\mathbb{N}$. That is, viewing $\mathbb{N}$ as an ordered set, its ordinal number is ω.

---

[2] The symbol ω, omega, is the last letter of the Greek alphabet.

Here are two other well-ordered sets:

$$\mathbb{N} \uplus \{z\} = \{0 < 1 < 2 < 3 < 4 < \cdots < z\} \quad \text{and}$$
$$\{z\} \uplus \mathbb{N} = \{z < 0 < 1 < 2 < 3 < 4 < \cdots\}.$$

As we previously noted, the first, $\mathbb{N} \uplus \{z\}$, is not order isomorphic to $\mathbb{N}$, and so it would be inappropriate to assign the value ω to it.

On the other hand, notice that $\{z\} \uplus \mathbb{N}$ is order isomorphic to $\mathbb{N}$:

$\mathbb{N}$	0	1	2	3	4	$\cdots$
	↓	↓	↓	↓	↓	
$\{z\} \uplus \mathbb{N}$	z	0	1	2	3	$\cdots$

Therefore, we assign the ordinal number ω to both $\{z\} \uplus \mathbb{N}$ and $\mathbb{N}$.

Ordinal numbers are also referred to as *order types* because they distinguish when a pair of well-ordered sets are isomorphic (same order type) or not (different order types).

Ordinal number arithmetic

Let A and B be disjoint, finite sets with a and b elements, respectively. It follows that

$$a + b = |A \cup B| \quad \text{and}$$
$$a \cdot b = |A \times B|.$$

Since cardinal and ordinal numbers coincide for finite sets, we likewise have that $A \uplus B$ is described by the ordinal number $a + b$ and $A \times B$ by $a \cdot b$.

We extend this to all ordinal numbers. Let α and β be ordinal numbers associated with the well-ordered sets A and B, respectively. We have the following definitions:

- $\alpha + \beta$ is the ordinal number associated with $A \uplus B$.

- $\alpha \cdot \beta$ is the ordinal number associated with $A \times B$.

Surprisingly, these operations are not commutative. Let's start with addition by comparing $1 + \omega$ and $\omega + 1$.

The ordinal $1 + \omega$ describes the ordered set $\{z\} \uplus \mathbb{N}$, which, as we observed, is order isomorphic to $\mathbb{N}$. Therefore, $1 + \omega = \omega$. By contrast, $\omega + 1$ describes the ordered set $\mathbb{N} \uplus \{z\}$, which is *not* order isomorphic to $\mathbb{N}$.

Conclusion: $1 + \omega \neq \omega + 1$.

> **Successors, well ordering, and the number before/after infinity**
>
> What's next? In a well-ordered set, it is possible to define a *successor* function. This is a function S that takes an element of the set as input and returns the next member of the set.
>
> Specifically, if x is an element of a well-ordered set A, then $S(x)$ is an element of A for which $x < S(x)$, and there is no element between them; that is, there is no element y with $x < y < S(x)$.
>
> For the natural numbers, $\mathbb{N}$, the successor of n is simply $n + 1$.
>
> In general, let x be an element of a well-ordered set A. It is possible that x is the largest element of A; in that case, x does not have a successor. Otherwise, let Z be the set of those elements of A that are larger than x. Since A is well ordered, the set Z has a least element, z. From this it follows that $x < z$, and there is no element of A between x and z; that is, z is the successor of x.
>
> We can also talk about the successor ordinal numbers. If α is the ordinal number describing a set A, then $\alpha + 1$ is the successor of α and describes the set $A \uplus \{a\}$.
>
> For the first infinite ordinal, ω, there is no ordinal number just before ω, but there is a number just after infinity: $\omega + 1$.

Next we show that multiplication is not commutative by comparing $2 \cdot \omega$ and $\omega \cdot 2$.

The ordinal number $2 \cdot \omega$ describes the ordered set $\{x < y\} \times \mathbb{N}$. This set contains all ordered pairs (x, n) and (y, n) where $n \in \mathbb{N}$. The lexicographical ordering of these pairs is:

$$(x, 0) < (x, 1) < (x, 2) < \cdots < (y, 0) < (y, 1) < (y, 2) < \cdots.$$

What we observe is a copy of $\mathbb{N}$ followed by another copy of $\mathbb{N}$. In other words:

$$2 \cdot \omega = \omega + \omega.$$

Now we consider $\omega \cdot 2$. This ordinal describes the ordered set $\mathbb{N} \times \{x < y\}$. This set contains all pairs (n, x) and (n, y) with $n \in \mathbb{N}$. Putting these in lexicographic order gives:

$$(0, x) < (0, y) < (1, x) < (1, y) < (2, x) < (2, y) < (3, x) < (3, y) < \cdots.$$

> **Same set, different orders, different ordinals**
>
> The cardinality of the natural numbers, $\mathbb{N}$, is $\aleph_0$. Using the usual less-than order, $\mathbb{N}$ is a well-ordered set corresponding to the ordinal number ω.
>
> However, we can imbue $\mathbb{N}$ with a different order that corresponds to a different ordinal number. For example, suppose we have all the even numbers in $\mathbb{N}$ (in their usual order) followed by all the odd numbers (also in their usual order), like this:
>
> $$0 < 2 < 4 < 6 < 8 < \cdots < 1 < 3 < 5 < 7 < \cdots.$$
>
> This is a different ordered set (albeit with the same elements) whose ordinal number is 2ω.

This ordered set is order isomorphic to $\mathbb{N}$, and therefore we have

$$\omega \cdot 2 = \omega$$

and we conclude that $2 \cdot \omega \neq \omega \cdot 2$.

We end this chapter with a brief look at ordinal exponentiation. The ordinal ω^2 is simply $\omega \cdot \omega$. More interesting is 2^ω.

We might think of this as an ordinal number describing this set:

$$X = \{0 < 1\} \times \{0 < 1\} \times \{0 < 1\} \times \{0 < 1\} \times \cdots.$$

This could be interpreted as the set of all infinitely long sequences $(a_0, a_1, a_2, \ldots)$ where each a_i is either a 0 or a 1, ordered lexicographically. However, X is not well ordered. Consider the subset of X containing these elements:

$$(1,1,1,1,1,1,\ldots) > (0,1,1,1,1,1,\ldots)$$
$$> (0,0,1,1,1,1,\ldots)$$
$$> (0,0,0,0,1,1,\ldots) > \cdots.$$

Notice that this subset has no least element, and therefore X is not well ordered.

Instead, we use a different ordered set, A, consisting of infinite sequences of 0s and 1s with the added requirement that each sequence has finitely many 1s. Imbuing A with lexicographic ordering makes A a well-ordered set, and 2^ω is its order type.

Generalizing this, we can consider the set B consisting of all sequences $(b_0, b_1, b_2, \ldots)$ where the b_js are natural numbers and only finitely many of the entries are nonzero. This set is also well ordered (by lexicographic ordering), and its order type is ω^ω.

This can get complicated very quickly. We can create well-ordered sets of type ω^{ω^ω} or worse:

$$\omega^{\omega^{\omega^{\cdot^{\cdot^{\cdot}}}}}$$

where the tower of exponents goes on forever!

A mind-boggling thought

There is a one-to-one correspondence between ordinal numbers and transfinite cardinal numbers. That is, each ordinal number α is associated with a cardinal number $\aleph_\alpha$. This implies that there is an equal quantity of ordinal numbers and transfinite cardinal numbers. This leads us to ask: How many are there?

Clearly, there are infinitely many ordinals starting with 0, 1, 2, 3, and then on to ω, $\omega+1$, $\omega+2$, and 2ω, 3ω, and ω^2, and many more. How many ordinal numbers are there? The formal way to frame this question is to consider the set Ω of all ordinal numbers and ask: What is the cardinality of Ω?[3]

The problem is: The set Ω doesn't exist! The proof is resonant of Russell's paradox that there can be no set of all sets. In a vague sense, there are too many sets for us to ask, How many sets are there? Likewise, there is no way to contain the ordinal numbers in a set to which we would assign a cardinality. They are beyond infinity.

For your consideration

In their usual order, the set of rational numbers, $\mathbb{Q}$, contains infinite decreasing sequences such as this:

$$1 > \frac{1}{2} > \frac{1}{3} > \frac{1}{4} > \frac{1}{5} > \cdots.$$

---

[3] Ω is the uppercase form of the Greek letter omega.

However, it is impossible to have an infinite decreasing sequence of natural numbers. Why?

Specifically, if A is a well-ordered set, explain why it is impossible to have an infinite decreasing sequence of elements of A.

Moreover, explain the converse: If B is an ordered set that does not contain an infinite decreasing sequence, then B must be well ordered.

Chapter 10

Infinite Shapes

> An infinite path is just another path; its only difference from other paths is that an infinite path arrives nowhere!
> —Mehmet Murat İldan, *Aforizmalar*

In this chapter we see how infinite processes lead to interesting mathematical shapes. We shall see how infinite paths, rather than arriving nowhere, arrive everywhere in our discussion of space-filling curves.

We begin with a familiar friend: the circle.

10.1 Circles

A *circle* is a plane figure consisting of all points a given distance (the radius) from a given point (the center). However, another way to think of a circle is as a regular polygon, but with infinitely many sides. This is how Archimedes found approximate values of π: the ratio of a circle's circumference to its diameter.

Start with a circle of radius 1; its circumference is 2π. An inscribed regular hexagon consists of six equilateral triangles (as shown in the left portion of Figure 10.1). The hexagon's perimeter is 6, which is smaller than the circumference of the circle. This implies that $6 < 2\pi$, or that $\pi > 3$.

Next, surround the circle with a regular hexagon (as shown in the right portion of Figure 10.1). The perimeter of this hexagon is greater than the circumference of the circle. Calculating the perimeter of this larger hexagon is a bit more involved.

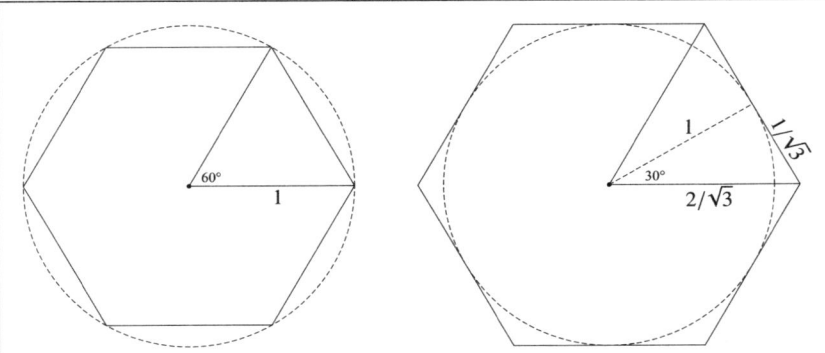

Figure 10.1: A circle of radius 1 with an inscribed hexagon (left) and circumscribed hexagon (right). The inscribed hexagon's perimeter is 6, giving $\pi > 3$. The circumscribed hexagon's perimeter is $12/\sqrt{3}$, giving $\pi < 6/\sqrt{3} \approx 3.4641$.

As before, the hexagon is composed of six equilateral triangles. To calculate the side lengths of one of these triangles, notice that the length of the altitude is 1; let $2x$ be the length of the side of the equilateral triangle. In this way, the altitude cuts the triangle into a right triangle whose sides have lengths $2x$ (the hypotenuse), 1, and x. By the Pythagorean theorem,

$$(2x)^2 = x^2 + 1 \quad \Rightarrow \quad 3x^2 = 1 \quad \Rightarrow \quad x = \frac{1}{\sqrt{3}}.$$

From this it follows that the perimeter of the hexagon is $6 \cdot 2/\sqrt{3} = 12/\sqrt{3}$. This gives

$$2\pi < \frac{12}{\sqrt{3}} \quad \Rightarrow \quad \pi < \frac{6}{\sqrt{3}} \approx 3.4641.$$

This analysis puts the value of π between 3 and 3.5. By increasing the number of sides, we achieve better approximations. Here is a table of lower and upper bounds for the value of π based on approximating a circle with inscribed and circumscribed regular polygons.

> **Calculating π using a formula by Ramanujan**
>
> While approximating π by calculating the perimeters of n-gons gives correct results, there are much more efficient methods. For example, one can use the following formula due to Srinivasa Ramanujan:
>
> $$\frac{1}{\pi} = \frac{2\sqrt{2}}{9801} \sum_{k=0}^{\infty} \frac{(4k)!(1103 + 26390k)}{(k!)^4 396^{4k}}.$$
>
> If we evaluate this sum merely up to $k = 4$, we get
>
> $$\pi \approx \frac{2262758404981105959754871105894245931679744}{509299577881529611662930757403081523769055\sqrt{2}}$$
>
> $$= 3.141592653589793\underline{0}2370938\ldots$$
>
> where the underlined 0 is the first incorrect digit.

Number of sides	Inscribed	Circumscribed
6	3.00000	3.46410
12	3.10583	3.21539
50	3.13953	3.14573
200	3.14146	3.14185
1000	3.14159	3.14160

10.2 How long is the diagonal of a square?

Regular n-gons look like circles when n is a large number. As n goes to infinity, the shape morphs from a polygon with pointy corners to an entirely smooth circle.

Archimedes' method of estimating π is valid because it finds lower bounds (from the inscribed polygon) and upper bounds (from the circumscribed polygon). We can ascertain that the value of π lies between those bounds.

Here we consider another situation for estimating a length in which we only have an upper bound.

> **Infinite sequences of shapes**
>
> What does it mean to let the number of sides of a regular polygon "go to infinity"? The situation is exactly like the infinite decimal 0.3333.... One can think of 0.3333... as an infinite sum, but it's better to think of it as a sequence whose terms can be made as close to $\frac{1}{3}$ as we like (see the presentation in Chapter 2).
>
> In a similar manner, when we say "let the number of sides go to infinity," it is better to think of a sequence of regular n-gons for n equal to 3, 4, 5, and so forth, like this:
>
>
>
> Letting the number of sides "go to infinity" actually means finding a shape that is approached by this sequence. To make that precise, we would develop a specific way to measure the difference between two shapes, and show that the "distance" between a regular n-gon and a circle can be made as small as we like.

A 1×1 square's diagonal has length $\sqrt{2}$; this is readily verified by using the Pythagorean theorem.

The idea is to approximate the diagonal of the square using a zig-zag path that looks like a staircase, as illustrated in Figure 10.2. Imagine that the diagonal of the square is approximated by an n-step staircase. Each step has a horizontal segment and a vertical segment; the lengths of these segments are all $\frac{1}{n}$. As n goes to infinity, the staircase morphs to become the diagonal of the square. But what of the lengths?

Surprisingly, the total length of an n-step staircase is 2 regardless of how tiny we make the steps. Of course, $\sqrt{2} \neq 2$, so the fact that the staircase approaches the diagonal as n tends to infinity doesn't mean that $\sqrt{2} = 2$.

What we can say is that the length of the staircase is an *upper bound* for $\sqrt{2}$ because the portion of the diagonal in each step is shorter than the length of the right-angle shaped step. All we may conclude is $\sqrt{2} \leq 2$.

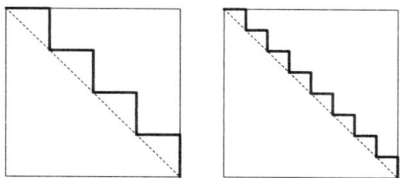

Figure 10.2: These images show how to approximate the diagonal of a 1 × 1 square with a staircase path. The length of the diagonal of the square is $\sqrt{2}$. As the number of steps in the path increases, we have a better and better approximation to the diagonal. However, in all cases, the total length of the zig-zag path is 2.

10.3 Sierpiński's carpet

One can think of a circle as the end result of a sequence of regular n-gons as n goes to infinity. Here we introduce a more exotic plane figure that was the brain child of the twentieth-century Polish mathematician Wacław Sierpiński.

We begin with a filled 1 × 1 square. Imagine this square as a 3 × 3 tic-tac-toe board of smaller squares. Remove the interior of the middle $\frac{1}{3} \times \frac{1}{3}$ square (upper left portion of Figure 10.3).

There are now eight $\frac{1}{3} \times \frac{1}{3}$ squares remaining after we punch out the interior of the middle square. The next step is to repeat this exact same process on those eight squares: remove the interior of the middle $\frac{1}{9} \times \frac{1}{9}$ square from each of the eight intact $\frac{1}{3} \times \frac{1}{3}$ squares (upper right portion of Figure 10.3).

At this point there are $8 \cdot 8 = 64$ intact squares. From each of these, punch out the interior of their inner $\frac{1}{27} \times \frac{1}{27}$ square (lower left portion of Figure 10.3).

Repeat this process over and over again. The resulting plane figure is known as *Sierpiński's carpet*. Let's think about its *area* and its *perimeter*.

Before we punch any holes, the original square's area is 1. Removing the middle $\frac{1}{3} \times \frac{1}{3}$ square removes $\frac{1}{9}$ of the area, leaving a total area of $\frac{8}{9}$.

The second step is exactly like the first. The eight remaining squares each lose one-ninth of their area. Stated differently, they retain exactly

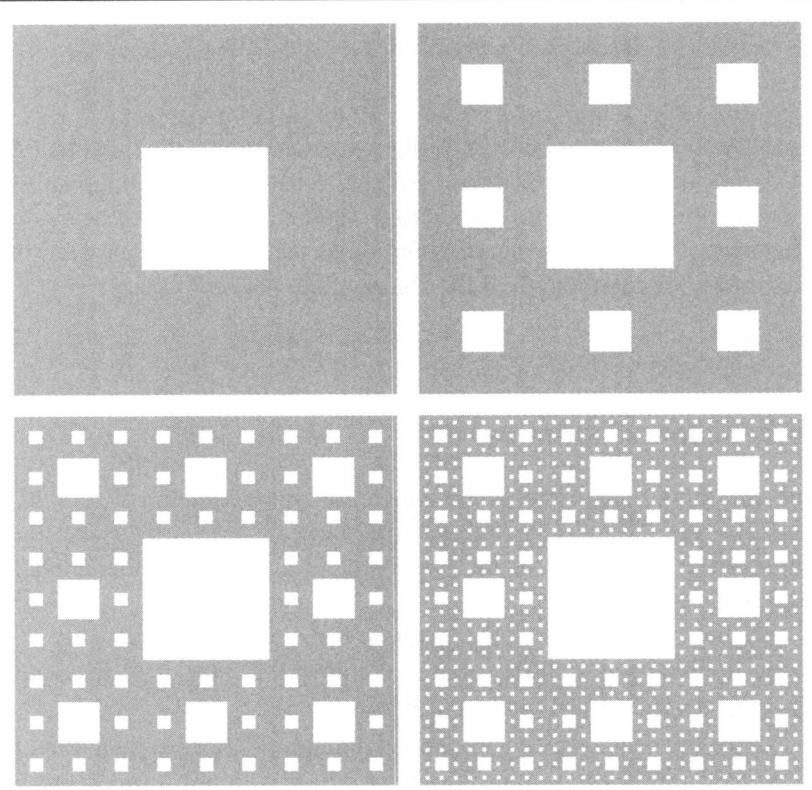

Figure 10.3: Sierpiński's carpet is constructed by punching holes in a 1×1 square. The first step is to remove the interior of the middle $\frac{1}{3} \times \frac{1}{3}$ square (upper left). This leaves eight $\frac{1}{3} \times \frac{1}{3}$ squares intact. The second step repeats this process on the interiors of the remaining eight $\frac{1}{3} \times \frac{1}{3}$ squares (upper right). This hole punching is done again (lower left) and again (lower right) *ad infinitum*.

$\frac{8}{9}$ths of their area. Since the area at the beginning of this step was $\frac{8}{9}$ and we keep only $\frac{8}{9}$ths of that, the amount of area remaining after this second step is $(\frac{8}{9})^2$.

At the start of the third step, the area is $(\frac{8}{9})^2$. As before, we remove one-ninth of the area from each of the 64 remaining little squares, leaving $\frac{8}{9}$ths of the total. Hence, at the end of step three, the total area is $(\frac{8}{9})^2 \cdot \frac{8}{9} = (\frac{8}{9})^3$.

It's now clear that after the completion of the nth step, the area remaining is $(\frac{8}{9})^n$. This implies that, as n goes to infinity, the total area of Sierpiński's carpet is zero!

This does not mean we have gobbled up the entire original unit square. At each stage of the process, we have removed the *interior* of the middle squares, not their boundary. Let's calculate the perimeter of Sierpiński's carpet: the total length of the edges left behind.

The original 1×1 square's perimeter is 4. When we punch a hole in the middle, the perimeter consists of both the original outer boundary and the boundary of the middle $\frac{1}{3} \times \frac{1}{3}$ square. (Remember, at each stage we are removing only the interiors of the middle squares.) Thus, at the end of the first step, the boundary's total length is $4 + \frac{4}{3} = \frac{16}{3}$.

In the second step, we punch out the middles of the eight $\frac{1}{3} \times \frac{1}{3}$ squares. The size of these middles is $\frac{1}{9} \times \frac{1}{9}$, so their perimeters are $\frac{4}{9}$. There are 8 of them, so they add $8 \cdot \frac{4}{9}$ to the boundary, for a new total of $4 + \frac{4}{3} + \frac{32}{9} = \frac{80}{9}$.

In the third step, the additional perimeter comes from the middles of the 64 remaining squares. These middles are $\frac{1}{27} \times \frac{1}{27}$ squares, so the added perimeter from the 64 new holes is $64 \cdot \frac{4}{27} = \frac{256}{27}$. At the end of step three, the total perimeter is

$$4 + \frac{4}{3} + \frac{32}{9} + \frac{256}{27}.$$

Let's do one more step and then generalize. At the start of step four, there are 512 intact squares whose middles are size $\frac{1}{81} \times \frac{1}{81}$. When we remove those interiors, each new hole adds $\frac{4}{81}$ to the perimeter, for a net addition of $512 \cdot \frac{4}{81} = \frac{2048}{81}$. At this point the perimeter is

$$4 + \frac{4}{3} + \frac{32}{9} + \frac{256}{27} + \frac{2048}{81}$$

which can be rewritten as

$$4 + \frac{4}{3} + \frac{8}{3} \cdot \frac{4}{3} + \left(\frac{8}{3}\right)^2 \cdot \frac{4}{3} + \left(\frac{8}{3}\right)^3 \cdot \frac{4}{3} = 4 + \frac{4}{3}\left[1 + \frac{8}{3} + \left(\frac{8}{3}\right)^2 + \left(\frac{8}{3}\right)^3\right].$$

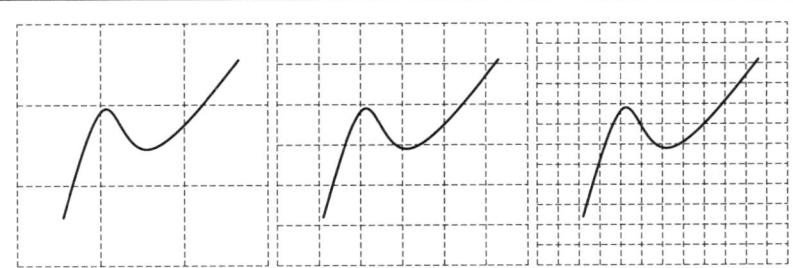

Figure 10.4: When a curve is overlaid with a grid of fineness f, the number of boxes touched by the curve is proportional to f.

The increasing powers of $\frac{8}{3}$ can be explained by the fact that after each step, the number of small squares removed increases eightfold, but the perimeters are one-third as large as previously. Thus, we find that the total perimeter of Sierpiński's carpet is

$$4 + \frac{4}{3}\left[1 + \frac{8}{3} + \left(\frac{8}{3}\right)^2 + \left(\frac{8}{3}\right)^3 + \cdots\right]. \qquad (10.1)$$

Notice that every term inside the brackets of (10.1) is greater than one, and we conclude that Sierpiński's carpet has infinite perimeter.

The dimension of Sierpiński's carpet

Familiar one-dimensional curves have finite length and zero area. Familiar two-dimensional plane shapes have finite perimeter and positive area. Is Sierpiński's carpet one-dimensional or two-dimensional? Amazingly, the answer is that its dimension is between those options.

There are various ways to define dimension; the method we employ involves overlaying figures with finer and finer grids.

Consider an ordinary curve in the plane, such as the one in Figure 10.4. On top of this curve, we draw a grid of vertical and horizontal lines. The *fineness* of the grid is a number f that gives the number of lines that span a unit distance. That is, if $f = 1$, the lines are distance 1 apart, but when $f = 2$, the lines are distance $\frac{1}{2}$ apart. In general, fineness f means that the distance between lines is $1/f$. Larger values of f are for finer meshes.

In the left portion of Figure 10.4, the curve sits on a grid with fineness $f = 1$. The curve intersects 5 boxes. In the middle portion, the fineness is increased to $f = 2$, and the same curve now intersects 10 boxes. Doubling the fineness to $f = 4$ results in the picture on the right, and now the curve passes through about 20 boxes. Were we to double the fineness again, each box would be subdivided into four smaller squares. On average, this doubles the number of boxes through which the curve passes. This estimate becomes more accurate as the grid gets finer. If N_f is the number of boxes intersected by the curve for a grid with fineness f, then the relationship between N_f and f can be written like this:

$$N_f \propto f \qquad (10.2)$$

where the $\propto$ symbol means *is proportional to*.

Let's do a similar analysis for a two-dimensional shape, such as the one shown in Figure 10.5.

In the left portion of the figure, the shape is overlaid with a grid with fineness $f = 1$. The middle has fineness $f = 2$, and the right has fineness $f = 4$. The box count increases with the fineness, but more rapidly. The majority of the boxes that overlap the shape are in its interior. Hence, when we double the fineness of the grid, the box count goes up by roughly a factor of 4. If we were to increase the fineness f by a factor of 10, each interior box would be subdivided a hundredfold, and N_f would be very nearly 100 times larger as a result.

For a two-dimensional figure, the relationship between N_f (the number of boxes intersecting the figure) and f (the fineness) can be expressed like this:

$$N_f \propto f^2. \qquad (10.3)$$

Equations (10.2) and (10.3) can be summarized as follows. For a d-dimensional figure, the number of boxes intersected by the figure, N_f, varies with the fineness, f, according to this relationship:

$$N_f \propto f^d. \qquad (10.4)$$

Let's apply this reasoning to Sierpiński's carpet. If we overlay it with a grid of fineness $f = 1$, then Sierpiński's carpet fits entirely in the unit square, so $N_1 = 1$. For some values of f, it is enormously difficult to calculate N_f exactly; however, when f is a power of 3, the situation is simple. In Figure 10.6, Sierpiński's carpet is overlaid with grids of

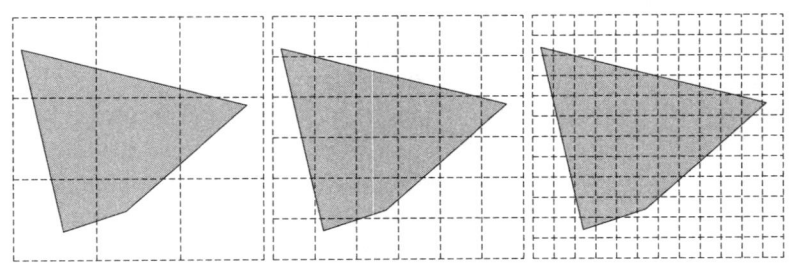

Figure 10.5: When a two-dimensional shape is overlaid with a grid of fineness f, the number of boxes touched by the shape is proportional to f^2.

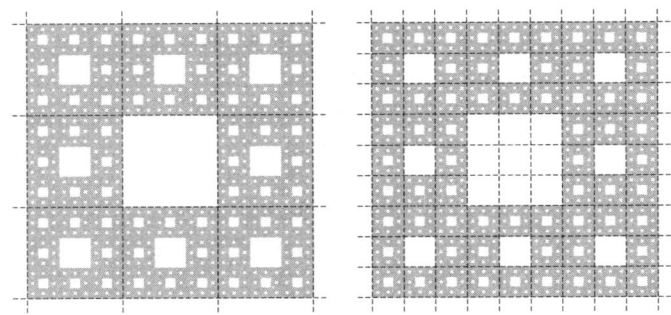

Figure 10.6: Sierpiński's carpet overlaid with a grid of fineness $f = 3$ (left) and with another of fineness $f = 9$ (right). We observe that $N_3 = 8$ and $N_9 = 64$.

fineness $f = 3$ and $f = 9$. Thanks to the structure of Sierpiński's carpet, it is easy to see that $N_3 = 8$ and that $N_9 = 64$.

Although not shown in the figure, increasing the fineness threefold to 27 results in another eightfold increase in N_f. We can capture this in an exact relationship:

$$N_{3^t} = 8^t. \tag{10.5}$$

The next step is to substitute $f = 3^t$ into equation (10.5). That makes the left-hand side simply N_f. To eliminate t from the right-hand side, we take logarithms (base 3) of both sides of (10.5):

$$\log_3 N_f = \log_3 8^t = t \log_3 8$$

and then exponentiate (base 3):

$$N_f = 3^{t \log_3 8} = \left(3^t\right)^{\log_3 8} = f^{\log_3 8}.$$

Let $d = \log_3 8$. We have found that $N_f = f^d$ when f is a power of three. With more care, one can show that this is more-or-less correct for any large value of f; we may write $N_f \propto f^d$ where $d = \log_3 8$.

Conclusion: The dimension of Sierpiński's carpet is $\log_3 8$, which evaluates to approximately 1.89.

The dimension of Sierpiński's carpet is a value that lies strictly between 1 and 2. Such shapes are known as *fractals* because their dimension is not a whole number.

10.4 Hilbert's space-filling curve

Sierpiński's carpet is created through an infinite, iterative process. We begin with a two-dimensional object (a filled square) and produce an object of lower dimension.

Here we do the opposite. We show how to transform a simple curve through an infinite, iterative process to entirely fill a square.[1] This curve was the creation of David Hilbert.

The process begins with the simple curve shown in Figure 10.7. This curve lies entirely inside the unit square whose lower left corner is at $(0, 0)$ and whose upper right corner is at $(1, 1)$. The curve proceeds through these coordinates:

$$\left(\tfrac{1}{4}, \tfrac{1}{4}\right) \to \left(\tfrac{1}{4}, \tfrac{3}{4}\right) \to \left(\tfrac{3}{4}, \tfrac{3}{4}\right) \to \left(\tfrac{3}{4}, \tfrac{1}{4}\right).$$

The fundamental step in the construction of Hilbert's curve follows these steps:

---

[1] The curves in this section are polygonal paths composed of a sequence of line segments. In other words, these curves are mostly straight with a lot of sharp 90° turns. The mathematical use of the word *curve* is broader than its common usage, which typically means a smooth contiguous path that has no straightaways.

152 ∞ Infinite Shapes

Figure 10.7: This three-segment path is the basic building block of Hilbert's space-filling curve. It lies in the interior of the unit square.

- Make four copies of the curve, each shrunk to half their original size.
- Place two of these copies along the top half of the unit square.
- Place the other two copies along the bottom half of the unit square, but rotate one a quarter turn clockwise (lower left) and rotate the other a quarter turn counterclockwise (lower right).
- Finally, the end of one piece of the curve is joined to the beginning of the next piece, as shown by the arrows in Figure 10.8.

Now we repeat this construction on the result, as illustrated in Figure 10.9.

We follow this basic recipe again and again. Further iterations are presented in Figure 10.10. When the process is allowed to run to infinity, the resulting curve passes through every point in the unit square!

A few technical notes

Hilbert's curve completely fills in the unit square, so in what sense is it a "curve"? To answer, let's start by giving a definition of what we mean by a curve in the plane.

A *curve* is a continuous function $f: [0, 1] \to \mathbb{R}^2$.

Let's unpack this. The set of inputs to this function is the unit interval: this is the set of all real numbers t with $0 \le t \le 1$. The outputs of the function lie in $\mathbb{R}^2$; that is, they are points in the plane.

It can be useful to think of the input t as *time* and the output as the location of a moving point as t goes from 0 to 1. The points returned by the function draw out the curve.

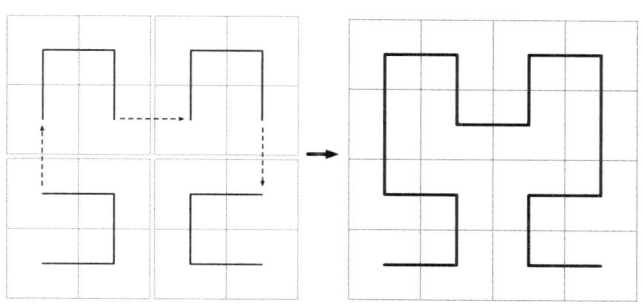

Figure 10.8: The first step in creating Hilbert's space-filling curve. Four half-size copies of the basic building block (see Figure 10.7) are arranged as shown. The top two copies are direct shrunken copies of the basic curve. The bottom two copies are rotated. The path is completed by connecting the end of one piece to the beginning of the next in the manner shown by the arrows. The end result is on the right.

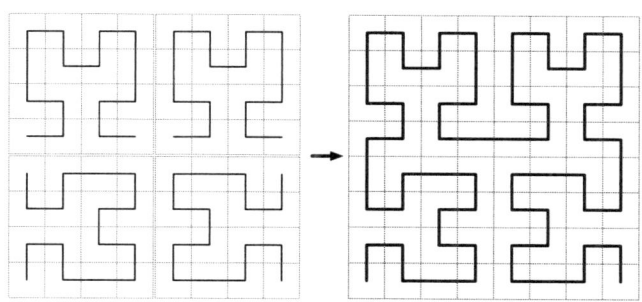

Figure 10.9: The second step in creating Hilbert's space-filling curve. Four half-size copies of the result of the first step (shown on the right in Figure 10.8) are placed and connected by the same method as in the first step.

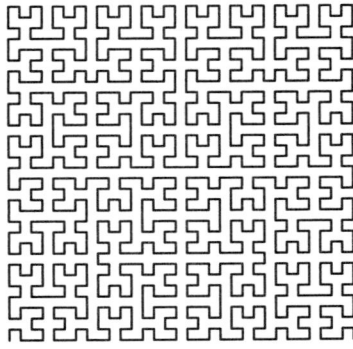

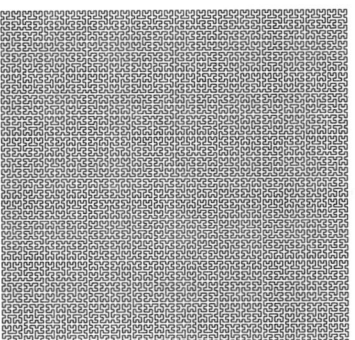

Figure 10.10: Further iterations in the creation of Hilbert's space-filling curve.

For example, define the function $c: [0, 1] \to \mathbb{R}^2$ by

$$c(t) = \bigl(\cos(2\pi t), \sin(2\pi t)\bigr).$$

As t goes from 0 to 1, the point $c(t)$ exactly traces a circle of radius one centered at the origin.

But isn't the *circle* the curve, not the function? Yes, we allow that mild ambiguity. We may speak interchangeably of the *function* $f: [0, 1] \to \mathbb{R}^2$ and the set of points that are the output of the function—the geometric figure traced out by f.

As mentioned earlier, it is acceptable for curves to have straight segments and sharp turns. For example, Hilbert's construction begins with

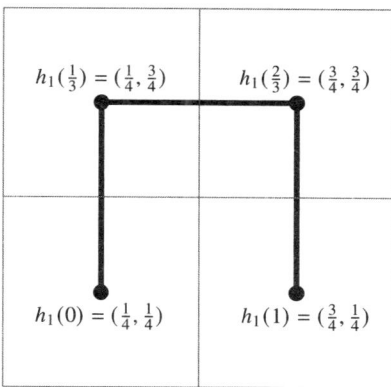

Figure 10.11: An annotated version of Figure 10.7 showing key values of the function $h_1:[0,1] \to \mathbb{R}^2$ that defines this curve.

the curve shown in Figure 10.7. Let's create a function $h_1:[0,1] \to \mathbb{R}^2$ that draws this curve.

Suppose the curve starts at the lower left. Then $h_1(0)$ should give the coordinates of that point: $(\frac{1}{4}, \frac{1}{4})$. The curve ends at the lower right, so we'd expect $h_1(1) = (\frac{3}{4}, \frac{1}{4})$.

The two corners, $(\frac{1}{4}, \frac{3}{4})$ and $(\frac{3}{4}, \frac{3}{4})$, lie at one-third and two-thirds of the length of the curve, so it makes sense to set $h_1(\frac{1}{3}) = (\frac{1}{4}, \frac{3}{4})$ and $h_1(\frac{2}{3}) = (\frac{3}{4}, \frac{3}{4})$ as shown in Figure 10.11.

What remains is to fill in the three segments for values of t in the gaps between $0, \frac{1}{3}, \frac{2}{3}$, and 1, like this:

$$h_1(t) = \begin{cases} (\frac{1}{4}, \frac{1}{4} + \frac{3}{2}t) & 0 \le t \le \frac{1}{3} \\ (-\frac{1}{4} + \frac{3}{2}t, \frac{3}{4}) & \frac{1}{3} \le t \le \frac{2}{3} \\ (\frac{3}{4}, \frac{7}{4} - \frac{3}{2}t) & \frac{2}{3} \le t \le 1. \end{cases} \qquad (10.6)$$

With a bit of work, we can check that substituting $t = 0, \frac{1}{3}, \frac{2}{3}, 1$ into h_1 yields the four corners, and that intermediate values of t smoothly move the output $h_1(t)$ along the line segments connecting those corners.

The next curve, depicted on the right in Figure 10.8, can be defined by a function $h_2\colon [0, 1] \to \mathbb{R}^2$. For this function, we have $h_2(0) = (\frac{1}{8}, \frac{1}{8})$ and $h_2(1) = (\frac{7}{8}, \frac{1}{8})$. Fully defining h_2, in a manner similar to the definition of h_1 in equation (10.6), is tedious, but conceptually the same.

In the same manner, we define h_3 to yield the curve in Figure 10.9, and then h_4, and so forth. It isn't necessary to write down messy formulas like the one in equation (10.6); it is enough to know that we can.

Given the functions $h_1, h_2, h_3, \ldots$, we can give a formula for Hilbert's curve. It is a function $h\colon [0, 1] \to \mathbb{R}^2$ defined like this:

$$h(t) = \lim_{n \to \infty} h_n(t). \qquad (10.7)$$

Important points remain:

- First, one must show that the limit in (10.7) exists.
- Second, one must show that the function so defined is continuous.
- Finally, one must show that for any point (x, y) in the unit square, there is a t such that $h(t) = (x, y)$.

These are not easy steps and are beyond the scope of this book.

10.5 Mandelbrot set

The Mandelbrot set is a fascinating shape of infinite complexity. It arises by considering a simple family of functions:

$$f_c(z) = z^2 + c$$

where c is a constant.

Iteration

Iteration is the act of repeatedly evaluating a function. That is, if f is a function and x is a number, we compute $f(x)$. Then we apply f again to that result, giving $f(f(x))$. Then we apply f to that result, and so forth.

A way to experiment with this is with a handheld calculator. Put any positive number into the calculator and press the square-root button. Then press that button again, and again, and again. What you are doing is iterating the square root function. If the first number you entered is 100, then you'll see a sequence of values like this:

$$100 \to 10.0 \to 3.162 \to 1.778 \to 1.334 \to 1.155 \to 1.075 \to \cdots.$$

Keep pressing the $\sqrt{}$ button, and eventually the calculator reaches 1.00000.

Next try iterating the cosine function. Enter any number into your calculator (be sure it is set to radians, not degrees), and repeatedly press the $\cos x$ button. Before long, the display settles in at 0.739085.

Repeatedly pressing the x^2 button has a rather different result. Start with any number greater than 1, and iteratively squaring takes the values off to infinity.

What happens when we iterate the function $f_c(z) = z^2 + c$ starting with $z = 0$?

If $c = 0$, the situation is rather boring. Since $f_0(z) = z^2 + 0$, the iterations starting at $z = 0$ are unchanging:

$$0 \to 0 \to 0 \to 0 \to 0 \to \cdots.$$

Let's look at iterations, starting from $x = 0$, with $c = 0.1$, $c = 0.2$, and $c = 0.3$.

- $f_{0.1}(z) = z^2 + 0.1$ gives these iterations starting from 0:

$$0 \to 0.1 \to 0.11 \to 0.1121 \to 0.11257 \to 0.11267$$
$$\to 0.11270 \to 0.11270 \to \cdots.$$

- $f_{0.2}(z) = z^2 + 0.2$ gives these iterations starting from 0:

$$0 \to 0.2 \to 0.24 \to 0.2576 \to 0.26636 \to 0.27095$$
$$\to 0.27341 \to 0.27475 \to 0.27549 \to 0.27589 \to 0.27589 \to \cdots.$$

- $f_{0.3}(z) = z^2 + 0.3$ gives these iterations starting from 0:

$$0 \to 0.3 \to 0.39 \to 0.4521 \to 0.50439 \to 0.55441 \to 0.60737$$
$$\to 0.66890 \to 0.74743 \to 0.85866 \to 1.03729 \to 1.37597$$
$$\to 2.19329 \to 5.11051 \to 26.41732 \to 698.17470 \to \cdots.$$

What we see is that for $c = 0.3$, the iterates of f_c, starting at 0, head off to infinity, but for $c = 0.1$ and $c = 0.2$, the iterates do not go to infinity; they converge. This is clearly visualized in the upper portion of Figure 10.12. The x-axis of this plot gives the iteration number and the y-axis gives the value of the iterates for each of the three functions $f_{0.1}$, $f_{0.2}$, and $f_{0.3}$.

Between 0.2 and 0.3, there is a change of behavior of the iterations of f_c. We tighten that window; the iterates of f_c for $c = 0.24$ and $c = 0.26$ are shown in the lower portion of Figure 10.12. Notice that with $c = 0.26$ the iterations tend to infinity, but not for $c = 0.24$.

The precise cutoff between the bounded and unbounded behaviors is $c = 0.25$. At this value, the iterates of $f_{0.25}$, starting from 0, increase toward but not past 0.5. For any larger value of c, the iterates of f_c tend to infinity.

Next let's explore some negative values for c and see how the iterates of f_c (always starting from 0) behave. The situation is more complicated. The iterations of $f_{-0.8}$ and $f_{-1.5}$ are presented in Figure 10.13. In both cases, the iterations bounce around, but they stay bounded—they do not go off to infinity.

The bounded behavior ends at $c = -2$. The iterates of f_c with $c = -1.999$ and $c = -2.001$ are plotted in Figure 10.14. With $c = -1.999$, the iterates bounce around chaotically but remain strictly between -2 and 2. However, with $c = 2.0001$, the iterations go off to infinity.

Let's summarize what we have observed when we iterate $f_c(z) = z^2 + c$ starting at $z = 0$. When c lies in the interval from -2 to 0.25, the iterations are bounded. However, if $c < -2$ or $c > 0.25$, then the iterations fly off to infinity.

Complex c

For real values of c, the behavior of the iterates of f_c is real simple: bounded when $-2 \le c \le \frac{1}{4}$ and off to infinity otherwise. For complex values of c, the situation is complex indeed. (All puns intended.)

Let's start with something simple: $c = i$. In this case, the iterations of f_i, starting at 0, have a simple pattern:

$$0 \to i \to -1+i \to -i \to -1+i \to -i \to -1+i \to -i \to \cdots.$$

The iterates simply oscillate between two values, but don't fly off to infinity.

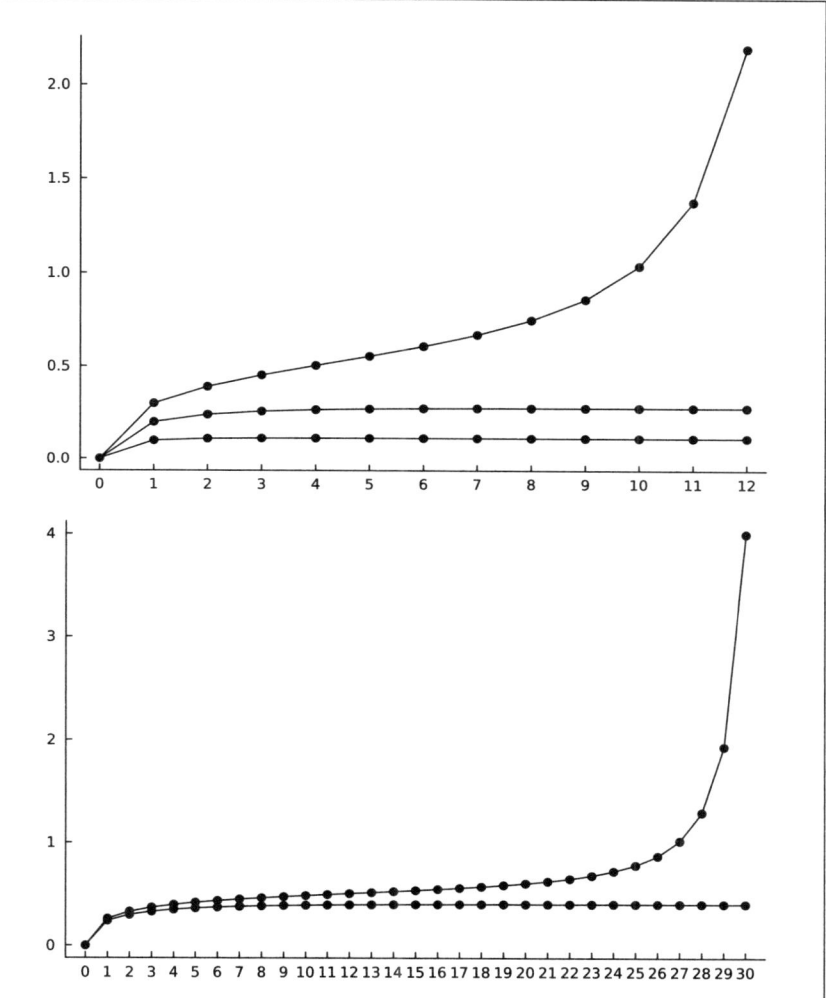

Figure 10.12: The upper portion of this figure shows iterations of the function $f_c(z) = z^2 + c$ starting at $z = 0$. The three plots correspond to $c = 0.1$ (bottom), $c = 0.2$ (middle), and $c = 0.3$ (top). The iterates of $f_{0.3}$ head off to infinity.

The lower portion shows iterations of f_c with $c = 0.24$ and $c = 0.26$.

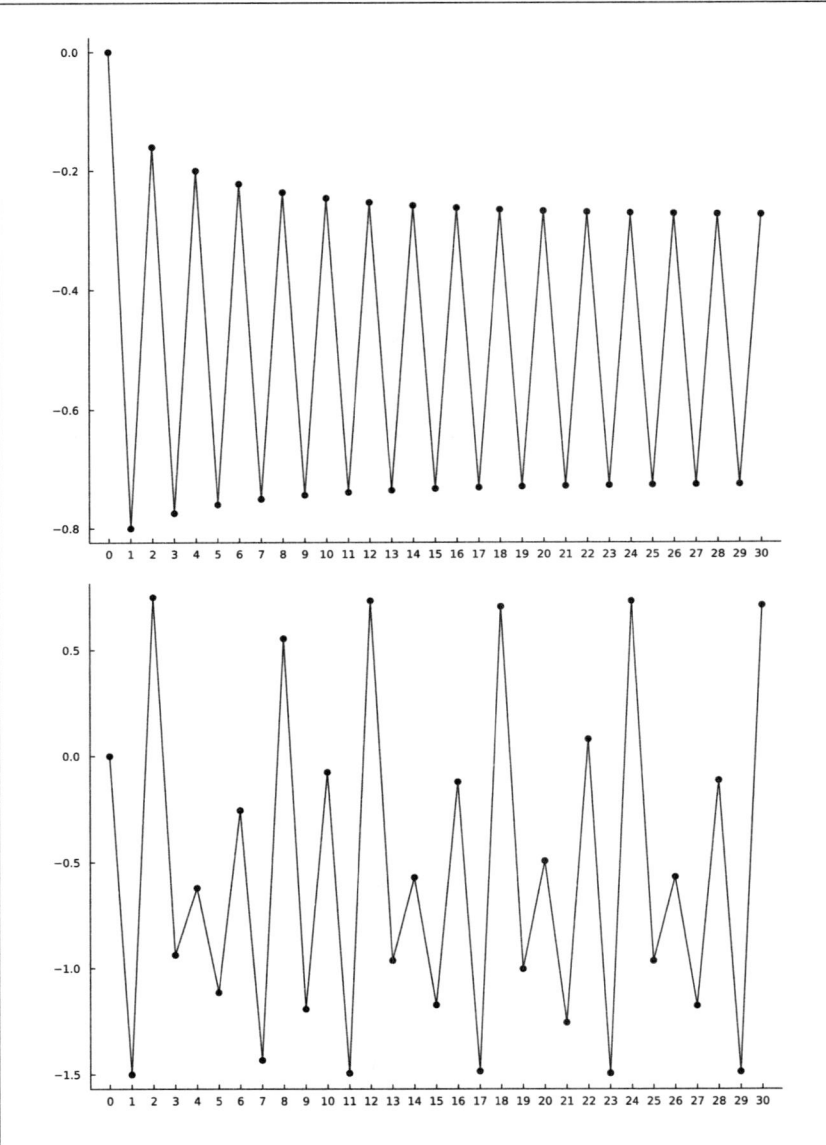

Figure 10.13: Iterations of f_c for $c = -0.8$ (upper) and $c = -1.5$ (lower).

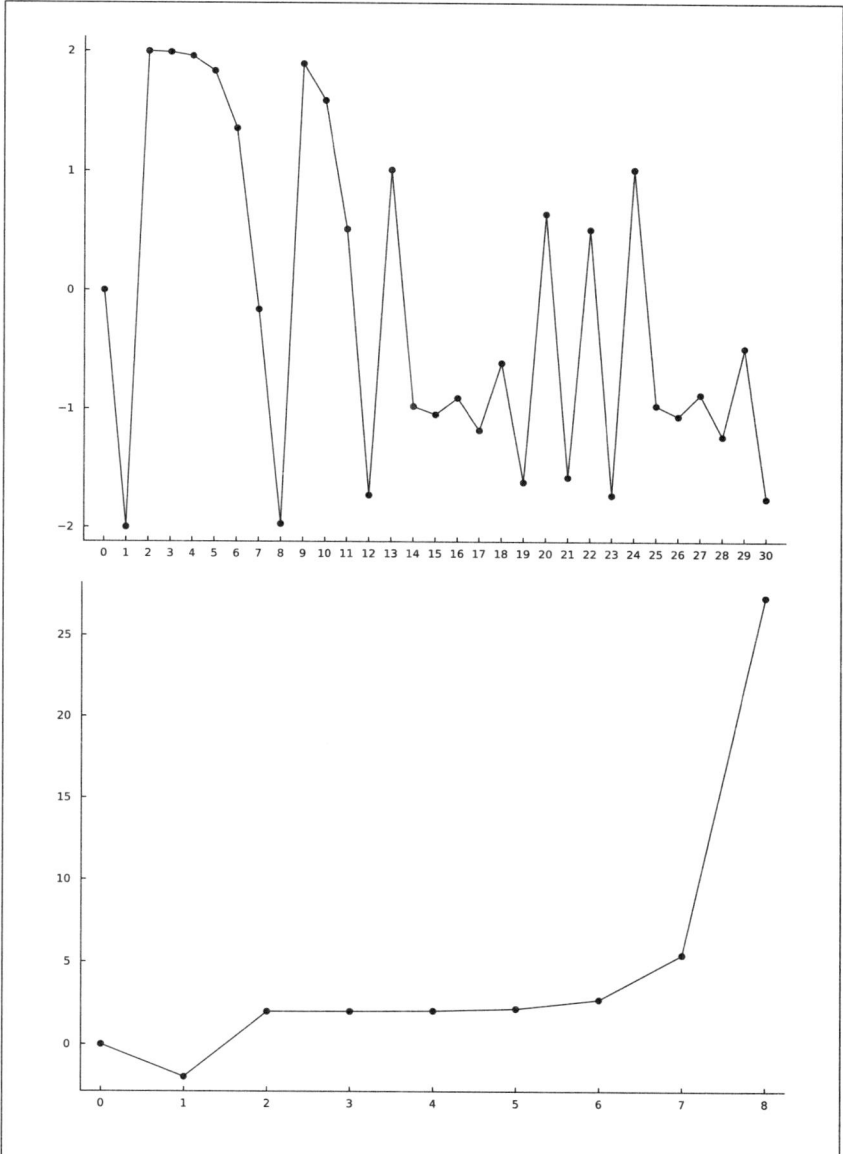

Figure 10.14: Iterations of f_c for $c = -1.999$ (upper) and $c = -2.001$ (lower).

Let's try another example: $c = 1 + i$. In this case, the iterates of f_c proceed as follows:

$$0 \to 1+i \to 1+3i \to -7+7i \to 1-97i$$
$$\to -9407 - 193i \to 88454401 + 3631103i \to \cdots.$$

Clearly, the iterations are heading off to infinity.

This leads to the question: For which values of c do the iterates of f_c stay bounded and for which values do the iterates go to infinity?

The *Mandelbrot set*, $\mathcal{M}$, named for Benoit Mandelbrot, consists of all complex numbers c such that the iterates of f_c, starting at 0, stay bounded. Conversely, if c is a complex number that is not an element of $\mathcal{M}$, then the iterations of f_c tend to infinity.

From our discussion so far, we know that $x + 0i$ is in $\mathcal{M}$ whenever $-2 \le x \le \frac{1}{4}$. We also have seen that i is an element of $\mathcal{M}$, but $1 + i$ is not.

The Mandelbrot set is a subset of the complex numbers, $\mathbb{C}$. As in Chapter 5, we can visualize $\mathbb{C}$ as a plane with the complex number $x + yi$ placed at coordinates (x, y); see Figure 5.1.

In this way, we can visualize $\mathcal{M}$ by coloring the point (x, y) black whenever $x + yi$ is in $\mathcal{M}$. The remarkable result appears in Figure 10.15.

The Mandelbrot set is a wonder of infinite complexity and beauty. Some parts of $\mathcal{M}$ are incredibly thin, so it's difficult to see that, in fact, $\mathcal{M}$ fills a contiguous region of the complex plane. Zooming in on portions of $\mathcal{M}$ reveals amazing patterns with exquisite detail. Two such magnifications can be seen in Figures 10.16 and 10.17.

It is easy to find software online to draw pictures of the Mandelbrot set on one's computer. These programs allow the user to zoom in on any portion of $\mathcal{M}$ to reveal amazing structures. They also add color to the points c not in $\mathcal{M}$ based on the speed with which the iterates go to infinity.[2]

For your consideration

Cantor's set is one of the most basic fractals. Its construction is a one-dimensional analog of Sierpiński's carpet.

We begin with the unit interval $[0, 1] = \{x \in \mathbb{R} : 0 \le x \le 1\}$. From this, we delete the interior of the middle third. That is, we delete all the

---

[2] Images and more information can be found on Wikipedia: en.wikipedia.org/wiki/Mandelbrot_set.

Figure 10.15: An overview of the Mandelbrot set, $\mathcal{M}$.

numbers x with $\frac{1}{3} < x < \frac{2}{3}$. The result is now the union of two intervals, each of length $\frac{1}{3}$, specifically, $[0, \frac{1}{3}] \cup [\frac{2}{3}, 1]$.

We now repeat this process on each of the two intervals. We remove the interior of the middle third of each of these; what remains is the union of four intervals, each of length $\frac{1}{9}$:

$$[0, \tfrac{1}{9}] \cup [\tfrac{2}{9}, \tfrac{1}{3}] \cup [\tfrac{2}{3}, \tfrac{7}{9}] \cup [\tfrac{8}{9}, 1].$$

The next step is to remove the interior of the middle third of each of these intervals, resulting in a set that is the union of eight intervals of length $\frac{1}{27}$. These steps are illustrated in Figure 10.18.

Cantor's set is the result of doing this process *ad infinitum*.

The total length of the "surviving" points in Cantor's set is calculated as follows. The first step removes $\frac{1}{3}$, which leaves a pair of intervals whose total length is $\frac{2}{3}$. The next step removes a third of that, leaving four

Figure 10.16: A close-up of the Mandelbrot set, $\mathcal{M}$, near the point $c = -0.748 + 0.0725i$.

intervals of total length $\frac{4}{9}$. The next step leaves eight intervals with total length $\frac{8}{27}$. In general, the total length after n steps is $\left(\frac{2}{3}\right)^n$, which tends to 0 as n goes to infinity.

Here are some items for your consideration.

To begin, using the box-counting idea (but restricted to one dimension), determine the dimension of Cantor's set. You should find that it is less than 1, but greater than 0.

Next, there are infinitely many points in Cantor's set; how many are there? Is it $\aleph_0$ or some other transfinite cardinal? Here's some help in figuring this out.

Figure 10.17: A close-up of the Mandelbrot set, $\mathcal{M}$, near the point $c = 0.29 + 0.161i$.

First, work in base three (ternary). Notice that all the numbers removed from $[0, 1]$ in the first step are strictly between 0.1_{THREE} and 0.2_{THREE}. Further, 0.1_{THREE} can also be written as $0.02222\ldots_{\text{THREE}}$. Let's adopt as a convention that if the ternary representation of a number ends in a 1, replace that with a 0 followed by infinitely many 2s.

The upshot of all this is that the numbers removed from $[0, 1]$ in the first step of the construction of Cantor's set have a 1 in the position just to the right of the ternary point.[3]

---

[3]We call it a *ternary point* because we are working in base three. It's only called a decimal point when working in base ten. The generic name for this symbol is *radix point*.

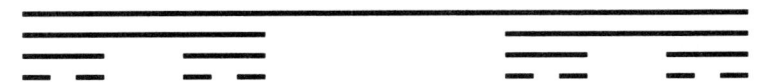

Figure 10.18: The first three steps in the construction of Cantor's set.

In the next step in the construction of Cantor's set, we remove all numbers x for which either $0 < x < \frac{1}{3}$ or $\frac{2}{3} < x < 1$. Let's write that in ternary:

$$0.0_{\text{THREE}} < x < 0.0222\ldots_{\text{THREE}} \quad \text{or} \quad 0.2_{\text{THREE}} < x < 0.2222\ldots_{\text{THREE}}$$

Look carefully and you will notice that the deleted numbers all have a 1 in the second position after the ternary point. The "surviving" numbers after the first two deletions are of the form $0.xy\ldots_{\text{THREE}}$ where the digits x and y are either 0 or 2.

After another step, all numbers with a 1 in the third position right of the ternary point are removed, and so forth. In the end, a number in the unit interval is a member of Cantor's set if and only if it can be written in ternary entirely with 0s and 2s.

For example, $0.20202020\ldots_{\text{THREE}}$ is in Cantor's set. (How would you write that as a simple fraction?)

Final step: Find a one-to-one correspondence between the elements of Cantor's set and the full unit interval $[0, 1]$. What does that tell you about the cardinality of Cantor's set?

Bibliography

Amir D. Aczel. *The Mystery of the Aleph: Mathematics, the Kabbalah, and the Search for Infinity*. Pocket Books, 2000.

James W. Anderson. *Hyperbolic Geometry*. Springer Undergraduate Mathematics Series. Springer, second edition, 2005.

Michael F. Barnsley. *Fractals Everywhere*. Dover, 2012.

Joseph Warren Dauben. *Georg Cantor: His Mathematics and Philosophy of the Infinite*. Princeton University Press, 1990.

Marc Diener and Francine Diener. *Nonstandard Analysis in Practice*. Universitext. Springer, 1995.

Amal El-Mohtar and Max Gladstone. *This Is How You Lose the Time War*. Saga Press, 2019.

Gustave Flaubert. *Madame Bovary*. Michel Lévy Frères, 1857.

Robert Goldblatt. *Lectures on the Hyperreals: An Introduction to Nonstandard Analysis*. Graduate Texts in Mathematics 188. Springer, 1998.

John Green. *The Fault in Our Stars*. Penguin Books, 2014.

Michael Hallett. *Cantorian Set Theory and Limitation of Size*. Clarendon Press, 1984.

Jonathan Halperin and Drew Takahashi (producers). *A Trip to Infinity*. Netflix, 2022.

Mehmet Murat İldan. *Aforizmalar*. Siyah Beyaz, 2022.

Claire Irving. Making the real projective plane. *The Mathematical Gazette*, 89(516):417–423, 2005.

Birger Iversen. *Hyperbolic Geometry*. London Mathematical Society Student Texts 25. Cambridge University Press, 1992.

James Joyce. *A Portrait of the Artist as a Young Man*. B. W. Huebsch, 1916.

Norton Juster. *The Phantom Tollbooth*. Bullseye Books, 1988.

Eric Katz. What is tropical geometry? *Notices of the American Mathematical Society*, 64(4):380–382, 2017.

Donald Knuth. *Surreal Numbers*. Addison-Wesley Professional, 1974.

Diane Maclagan and Bernd Sturmfels. *Introduction to Tropical Geometry*. American Mathematical Society, 2015.

Cindy Nemser. An interview with Eva Hesse. *Artforum*, 8(9):59–63, May 1970.

Edward Olney. *General Geometry and Calculus*. Sheldon and Company, 1871.

Edgar A. Poe. *Eureka: A Prose Poem*. Wiley & Putnam, 1848.

Arlan Ramsay and Robert D. Richtmyer. *Introduction to Hyperbolic Geometry*. Universitext. Springer-Verlag, 1995.

Alain Robert. *Nonstandard Analysis*. Wiley, 1985.

Abraham Robinson. *Non-standard Analysis*. Princeton University Press, subsequent edition, 1996.

Rudy Rucker. *Infinity and the Mind: The Science and Philosophy of the Infinite*. Paladin, 1984.

Edward Scheinerman. *The Mathematics Lover's Companion*. Yale University Press, 2021.

Edward Scheinerman. *From Counting to Continuum: What Are Real Numbers, Really?* Cambridge University Press, 2024.

Charles Seife. *Zero: The Biography of a Dangerous Idea*. Penguin Books, 2000.

David Speyer and Bernd Sturmfels. Tropical mathematics. *Mathematics Magazine*, 82(3):163–173, 2009.

Ian Stewart. *Infinity: A Very Short Introduction.* Oxford University Press, 2017.

Index

Δ, 41
Ω, 139
$\Rightarrow$, 39
$\aleph_0$, 119
$\odot$, 24
ω, 135
$\oplus$, 24
$<$, 127
$\propto$, 149
$\times$, 132
⊎, 131
$\vee$, 39
$\wedge$, 39

$\forall$, 38
addition, tropical, 24
additive inverse, 6
algebraic curve, 28
algebraic number, 123
angular defect, 99
antipodal, 80
Archimedes, 141
associative property, 5, 25
Axiom of Choice, 134

Bézout's theorem, 33, 74
bijection, 107–110

$\mathfrak{c}$, 120
$\mathbb{C}$, 47

$\overline{\mathbb{C}}$, 50
Cantor, Georg, 114
Cantor-Schröder-Bernstein theorem, 118
Cantor's set, 162
cardinal, transfinite, 118–121
cardinality, 105
Cartesian product, 132
Cohen, Paul, 120
commutative property, 5, 25
complex numbers
    extended, 50
complex plane, 49
complex projective plane, 79
concurrent, 67
continuum, 120
continuum hypothesis, 120
    generalized, 122
converge, 19
coordinates
    homogeneous, 65, 71
    polar, 53
countable, 119
curve, 151, 152
    algebraic, 28

de Man, Cornelis, 61
Desargues, Girard, 67
Desargues's theorem, 67–69

Diller, Phyllis, 23
distributive property, 5, 25
duality, 62, 67

∃, 39
El-Mohtar, Amal, 35
element
    identity, 5, 26
empty set, 110
Escher, M. C., 97
Euclidean plane, 60
existential quantifier, 39
extended
    complex numbers, 50
    real numbers, 8

Fano plane, 80
Fault in Our Stars, The, 103
field, ordered, 38
first-order statement, 40
Flaubert, Gustave, 83
fractal, 151
Fraenkel, Abraham, 121
function, 105–106
    successor, 137
Fundamental Theorem of Algebra, 75

Gladstone, Max, 35
Green, John, 103

harmonic series, 20
height, 124
Heron's formula, 99
Hesse, Eva, 59
Hilbert
    curve, 151–156
    hotel, 112–113
Hilbert, David, 151
Hofstadter, Douglas, 127

homogeneous, 64
    coordinates, 65, 71
hotel, Hilbert, 112–113
hyperreal numbers, 35

ideal point, 86
identity element, 5, 26
IEEE Standard 754, 11
İldan, Mehmet Murat, 141
inverse
    additive, 6
    multiplicative, 6, 26

Joyce, James, xiii
Juster, Norton, 15

lemniscate, xv
lexicographic order, 132
line at infinity, 61
linear fractional transformation, 53, 92

$\mathcal{M}$, 162
Madame Bovary, 83
Man Weighing Gold, A, 61
Mandelbrot, Benoit, 162
Mandelbrot set, 162
meta-theorem, 40
min-sum arithmetic, 24
Möbius band, 73
multiplication, tropical, 24
multiplicative inverse, 6, 26

nan, 12
natural number, 8
negative, 6, 7
number
    algebraic, 123
    cardinal, 118–121
    hyperreal, 35

Index ∞ 173

natural, 8
ordinal, 133–139
real, 4
transcendental, 123–124

one-to-one, 107
onto, 107
order
   isomorphic, 129
   lexicographic, 132
   preserving, 129
   relation, 128
   type, 136
ordered
   field, 38
   set, 127
ordinal number, 133–139

$\mathscr{P}$, 114
parallel postulate, 83
Phantom Tollbooth, The, 15
pigeon-hole principle, 108
place value notation, 15
Playfair, John, 84
Poe, Edgar Allan, xi
Poincaré, Henri, 86
point at infinity, 61
polar coordinates, 53
Portrait of the Artist as a Young Man, A, xiii
positive, 7
power set, 114, 122
projective plane, 61
   complex, 79
property
   associative, 5, 25
   commutative, 5, 25
   distributive, 5, 25
   transitive, 6, 128
   trichotomy, 6, 128

Pythagorean theorem, 97

quantifier, 38, 39

$\mathbb{R}$, 4
$\overline{\mathbb{R}}$, 8
$^*\mathbb{R}$, 35
radix point, 165
Ramanujan, Srinivasa, 143
real numbers, 4
   extended, 8
reciprocal, 6
Riemann, Bernhard, 51
Riemann sphere, 50–51
Robinson, Abraham, 35
Russell, Bertrand, 121
Russell's paradox, 121, 139

set, 104–105
   empty, 110
   ordered, 127
   power, 114, 122
   well-ordered, 129–133
Sierpiński, Wacław, 145
Sierpiński's carpet, 145–151
size, 105
slope, 41, 61
space-filling curve, 151–156
standard part (st), 37, 42
stereographic projection, 50
successor function, 137
surreal numbers, 41

$\mathbb{T}$, 24
ternary, 165
theorem
   Bézout, 33, 74
   Cantor-Schröder-Bernstein, 118
   Desargues's, 67–69
   Fundamental, of Algebra, 75

meta, 40
Pythagorean, 97
This Is How You Lose the Time War, 35
topology, 70
transcendental number, 123–124
transfer principle, 37–40
transfinite cardinal, 118–121
transitive property, 6, 128
trichotomy property, 6, 128

uncountable, 120
undefined, 6
universal quantifier, 38
upper bound, 44
upper half-plane model, 91

von Neumann, John, xii

well-ordered set, 129–133
Wright, Steven, 3, 47

Zermelo, Ernst, 121
ZF axioms, 121, 134